"The philosophical significance of discovering life beyond Earth i[...] immense. This book is a wonderful advancement of current discourse [...] philosophical implications of extraterrestrial life."

Andrew M. Davis, *The Center for Process Studies,*
Claremont School of Theology, USA

EXOPHILOSOPHY

This volume addresses philosophical questions raised by the possibility of alien life and extraterrestrial intelligence. The different philosophical perspectives and approaches presented across the chapters will provide a foundation for future work on exophilosophy.

Interest in space, space exploration, and alien life has never been greater. In popular culture, for example, it has proven a persistent theme in science fiction films (e.g., *Star Trek*, *Star Wars*), books (e.g., H. G. Wells, Arthur C. Clarke, Ray Bradbury), and computer games (e.g., Sid Meier's *Alpha Centauri*), as well as bestselling 'non-fiction' books (von Däniken's multimillion-selling *Chariots of the Gods?*), and hit 'documentary' shows (e.g., *Ancient Aliens*). There has also been persistent interest in these topics amongst scientists with organizations such as NASA and SETI having an enormous impact on both the scientific and popular imagination. Yet, curiously, the topic has received relatively little philosophical attention. Whilst certain aspects of these topics remain within the proper purview of the sciences, a host of philosophical questions are raised by the possibility of alien life and extraterrestrial intelligences, and the possibility of our coming into contact with them. This collection of essays will examine some of these questions whilst laying the groundwork for future study in an as-yet under-researched area of philosophy.

Exophilosophy is essential reading for scholars and students with an interest in space and philosophy, especially those working in philosophy of science, metaphysics, epistemology, ethics, philosophy of language, and philosophy of religion.

Richard Playford is currently Senior Lecturer in Philosophy, Ethics, and Religion at Leeds Trinity University. He holds a PhD from the University of Reading. His most recent publication, prior to this, co-authored with Stephen Bullivant and Janet Siefert, is *God and Astrobiology* (2024).

EXOPHILOSOPHY

The Philosophical Implications of Alien Life

Edited by Richard Playford

Routledge
Taylor & Francis Group

NEW YORK AND LONDON

Cover image via Getty: Han Jaihui

First published 2025
by Routledge
605 Third Avenue, New York, NY 10158

and by Routledge
4 Park Square, Milton Park, Abingdon, Oxon, OX14 4RN

Routledge is an imprint of the Taylor & Francis Group, an informa business

Library of Congress Cataloging-in-Publication Data
Names: Playford, Richard, editor.
Title: Exophilosophy : the philosophical implications of alien life /
edited by Richard Playford.
Description: New York, NY : Routledge, 2025. |
Includes bibliographical references and index.
Identifiers: LCCN 2024033803 (print) | LCCN 2024033804 (ebook) |
ISBN 9781032576084 (hardback) | ISBN 9781032576091 (paperback) |
ISBN 9781003440130 (ebook)
Subjects: LCSH: Extraterrestrial beings–Philosophy. | Life on other planets–Philosophy.
Classification: LCC QB54 .E93 2025 (print) | LCC QB54 (ebook) |
DDC 576.8/3901–dc23/eng/20240929
LC record available at https://lccn.loc.gov/2024033803
LC ebook record available at https://lccn.loc.gov/2024033804

ISBN: 978-1-032-57608-4 (hbk)
ISBN: 978-1-032-57609-1 (pbk)
ISBN: 978-1-003-44013-0 (ebk)

DOI: 10.4324/9781003440130

Typeset in Sabon
by Newgen Publishing UK

CONTENTS

CONTRIBUTORS

Stephen Bullivant is Professor of Theology and the Sociology of Religion at St Mary's University, UK, and Professorial Research Fellow in Theology and Sociology at the University of Notre Dame, Australia. His recent books include *God and Astrobiology* (2024; co-authored with R. Playford and J. Siefert) and *The Cambridge History of Atheism* (2021; co-edited with M. Ruse).

David Ellis is a Senior Lecturer in Philosophy, Ethics and Religion at Leeds Trinity University. His research applies Wittgenstein's later philosophical views to religion, science, and alien communication.

Chelsea Haramia is Associate Professor in the Department of Philosophy at Spring Hill College, USA. She is currently serving as Senior Research Fellow at the Center for Science and Thought at the University of Bonn, Germany, on a project titled *Desirable Digitalisation: Rethinking AI for Just and Sustainable Futures*. She also has a fellowship with the Academy of International Affairs at the University of Bonn to work on her project titled *Global Planning for Post-Detection*. She currently has a book under contract on the morality of cosmic messaging. Her research focuses on the intersections of science, technology, and values. She is the author of several articles and book chapters on space exploration, astrobiology ethics, and the search for extraterrestrial technology, as well as public philosophy news articles and outreach. She holds a PhD in philosophy and a graduate certificate in gender and women's studies from the University of Colorado, Boulder. She is co-editor of the journal *1000-Word Philosophy*, and an international affiliate of the UK SETI Research Network's Post-Detection Hub, an affiliate of the SETI Institute.

Lewis Howeth is currently pursuing a PhD in philosophy and religion. He holds a master's degree from Durham University and a bachelor's degree from Newcastle University. His research interests focus on animal ethics from a Catholic natural law perspective. His work in this area aims to bridge traditional religious ethical frameworks with contemporary issues in animal rights. In addition to his academic pursuits, he runs a YouTube channel (*Perspective Philosophy*) that has garnered over 1 million views. Through his channel, he has interviewed prominent figures, such as best-selling author and philosopher Peter Singer and David Pearce of the transhumanist movement, offering accessible insights into complex philosophical debates. Lewis has also shared his research and insights at prestigious institutions, including the Faculty of Divinity at Cambridge University and Trinity College Dublin.

Jeffery D. Long is the Carl W. Zeigler Professor of Religion, Philosophy, and Asian Studies at Elizabethtown College, in Pennsylvania, where he has taught since receiving his doctoral degree from the University of Chicago Divinity School in the year 2000. He has written many books and articles, including *Jainism: An Introduction* and *Discovering Indian Philosophy*. In 2022, his book, *Hinduism in America: A Convergence of Worlds*, received the Rajinder and Jyoti Gandhi Award for Excellence in Philosophy, Theology, and Critical Reflection. That year, he also received an award from the International Ahimsa Foundation, a Jain organization based in New York. He has spoken at the United Nations and appears in a PBS documentary on the life of Swami Vivekananda.

Thomas Metcalf is an associate professor of philosophy at Spring Hill College in Mobile, Alabama, USA. His main philosophical interests are in philosophy of science, philosophy of religion, and social and political philosophy. He is also a research at the Institute for Science and Ethics, affiliated with the University of Bonn, in Bonn, Germany.

Robert CB Miller is a former financial journalist and company director. He has a PhD in philosophy from the University of Reading. His research interests include virtue ethics, philosophy of mind, and apologetics.

Richard Playford is a Senior Lecturer in Philosophy, Ethics and Religion at Leeds Trinity University. He has a PhD in philosophy from the University of Reading, a master's degree in philosophy from the University of Birmingham, a bachelor's degree in philosophy with ancient history from the University of Exeter, and a postgraduate certificate in higher education from Leeds Trinity University. He is also a fellow of the Royal Society of Arts and a fellow of the Higher Education Academy. Generally, his research interests lie at the intersection of ethics and metaphysics, with a particular emphasis on the thought of Aristotle and Aquinas.

Gregory Stacey completed his DPhil in Theology at the University of Oxford and has lectured in philosophy and theology at several UK universities. He is currently Assistant Professor of Franciscan Theology at Saint Francis University, Pennsylvania. His research explores Catholic theology through the lens of analytic metaphysics and epistemology, and also examines the epistemology of musical experience.

INTRODUCTION

Richard Playford

Exophilosophy

One of the things that makes philosophy endlessly fascinating is that (almost) anything can be examined through its lens! One can take (almost) any subject, feature of the world, or facet of the human condition, add 'philosophy of ...' in front of it, and you now have a perfectly respectable field of philosophical enquiry. Examples include philosophy of language, philosophy of art, philosophy of science, philosophy of architecture, philosophy of history, philosophy of sex and gender, and so on. If we wished, we could continue to multiply examples (almost) endlessly; however, hopefully, my point is clear enough.

One particularly exciting possibility is that of alien life. Are we alone in the Universe? Is there life out there? If so, what is it like? Are there other intelligent life forms in the Universe? Could we ever hope to make contact with them or even visit them? Questions like these lie at the intersection of science and imagination. These questions can be, and have been, fruitfully examined from a scientific perspective, and we will go on to discuss how this has been done shortly. At the same time, these questions can be, and have been, fruitfully explored through a creative, speculative, and fictional lens. Aliens feature prominently in science fiction books, such as H. G. Wells's *The War of the Worlds*; television series, such as Gene Roddenberry's *Star Trek*; films, such as George Lucas's *Star Wars*; video games, such as Sid Meier's *Alpha Centauri*; and in many other mediums besides.

As a result, it should hardly surprise us that alien life can also be fruitfully examined through a philosophical lens. As we shall see, the possibility of alien life raises a number of fascinating philosophical questions. What might surprise us, however, is the relative scarcity of serious philosophical discussions of alien life. For example, searches for 'exophilosophy', 'astrophilosophy', 'extraterrestrial', 'extraterrestrials', and 'aliens' reveal no relevant

DOI: 10.4324/9781003440130-1

dedicated articles on either the *Stanford Encyclopedia of Philosophy* or the *Internet Encyclopedia of Philosophy*, two of the most prominent academic online philosophical encyclopedias. As far as I can see, these topics are simply not discussed, at least not as topics in and of themselves.

Thankfully, this is beginning to change.[1] Exophilosophy is beginning to grow as a distinct philosophical subdiscipline, yielding new insights into old questions whilst generating new avenues of enquiry and research. This collection of essays aims to add to this growing body of thought in an as yet under-researched philosophical subdiscipline.

Astrobiology

The topic of this volume is exophilosophy. Thus, whilst the focus of what follows is philosophical in nature, the topics are explored against a scientific backdrop, namely that of astrobiology. As a result, it is perhaps worth briefly considering astrobiology as a scientific discipline in order to better understand the backdrop against which the chapters of this volume are written. With that in mind, we now ask, what is astrobiology? How did it emerge as a distinct scientific subdiscipline, that is, as a distinct field of study with its own norms, methods, methodologies, and experts? And what do the latest findings in astrobiology tell us about the possibility of extraterrestrial life?

Astrobiology is a branch of science that studies the origins, evolution, distribution, and future of life in the Universe (Kaufman, 2022).[2] Whilst scientists and thinkers have considered the possibility of extraterrestrial life throughout the history of human thought, astrobiology as a distinct and formal scientific subdiscipline has its origins in the space race and the Cold War. In 1952, Stanley Miller and Harold Urey simulated the chemical conditions of the early Earth by mixing methane, ammonia, hydrogen, and water together. They then zapped this solution with an electrical charge in order to simulate lightning. A black goo resulted which, upon analysis, contained amino acids. Amino acids are one of the building blocks of cells and, thus, of life. Thus, Miller and Urey had demonstrated how some of the building blocks of life could have formed through entirely physical and chemical processes on the early Earth. Whilst this is a long way from demonstrating how life could have arisen through such processes (a question to which we still do not know the answer), this groundbreaking discovery moved us one step closer.[3] The results were published in the journal *Science* in 1953 (Miller, 1953).[4]

NASA embraced their results, and NASA's first exobiology research award went to Wolf Vishniac in 1959 for the 'Wolf Trap' Life Detector, which was designed to fly on a Viking lander (although due to budget cuts, it was not used during the actual mission in 1976). NASA then went on to establish an Exobiology Programme in 1970 (NASA, 2024). Throughout the 1970s, 1980s, and 1990s, the field widened, grew, and matured, and so in 1998 NASA established their Astrobiology Programme which incorporated the existing Exobiology Programme. Thus, exobiology with its narrower scope laid the groundwork for astrobiology which is broader in scope (Soffen, 1997). Of course, it wasn't only NASA that was interested in this topic, and over the years various other space agencies, private companies, and universities have started programmes and research into these areas. Thus, astrobiology as a distinct scientific subdiscipline was born!

THE DRAKE EQUATION

$$N = R* \times f_p \times n_e \times f_l \times f_i \times f_c \times L$$

N = the number of extraterrestrial civilizations with which human beings could communicate
$R*$ = average rate at which stars form in our galaxy.
f_p = fraction of stars with planets.
n_e = number of planets that could potentially support life per star that has planets.
f_l = fraction of planets that could potentially support life that actually go on to develop life.
f_i = fraction of life-supporting planets that develop intelligent life.
f_c = fraction of intelligent species that develop the capacity for interstellar communication.
L = average length of time that such communicative civilizations are active.

Alongside the Miller–Urey experiment, a number of other key discoveries and concepts are worth examining when trying to understand contemporary astrobiology. One of these is the Drake equation. In 1961 Frank Drake (1930–2022) created a formula to estimate the number of advanced communicative (via radio waves) extraterrestrial civilizations currently active in the Milky Way (our galaxy). The Drake equation involves a number of variables, only some of which are known. When Drake first formulated his equation, only the average rate at which stars form in our galaxy ($R*$) was known, with scientists more or less having to guess the answers to the other variables. Since then, we have made significant progress in determining the percentage of stars with planets (f_p), and some progress towards determining the number of planets that could potentially support life per star that has planets (n_e), although even today there is still a degree of uncertainty when attempting to precisely quantify these factors. The other factors remain basically unknown, and as a result, there is an enormous amount of variation when scientists try to estimate the number of extraterrestrial civilizations with which human beings could communicate (N) with estimates ranging from 1 (namely *us*, i.e., we're alone in the galaxy) to several million. Later in life, Drake himself put the estimate at around 10,000 (Shostak, 2021).

Some time later, Peter Ward and Donald E. Brownlee formulated the 'Rare Earth' hypothesis. In essence, they did this by adding a whole series of additional requirements to the Drake equation, the result of which is that the emergence of complex life is immensely rare and may have only happened once in the history of the galaxy or even the entire Universe (Ward and Brownlee, 2000).

Also worthy of note is the Fermi paradox, named after the Italian–American physicist Enrico Fermi (1901–1954).[5] The Fermi paradox asks why, given the enormous size of the Universe, we haven't yet discovered or been contacted by any other forms of intelligent life. The sheer size of the Universe would seem to imply that there ought to be other intelligent species out there, and yet the deafening silence of the Universe would seem to suggest we are alone, hence the paradox.[6] Numerous solutions to the Fermi paradox exist, including the previously mentioned Rare Earth hypothesis. Another possibility is the rather depressing idea that intelligent life tends to destroy itself eventually (Cai et al., 2021). Another solution is the terrifying 'Dark forest hypothesis', according to which it is possible that there is a highly advanced highly aggressive alien species out there which promptly attacks and destroys any other intelligent species it detects (Yu, 2015). As a result, whilst there may be many intelligent species out there, most of them are staying quiet so as to avoid being detected and killed! (If this is the case, we really ought to keep quiet too!) There is also the seemingly bizarre 'Zoo hypothesis', according to which more advanced species are keeping us isolated or are hiding from us for whatever reason (perhaps in order to study us, or to protect us, or for some other reason) (Forgan, 2016). Other solutions have been proposed. There is insufficient space to describe them all here, but hopefully, these examples illustrate how varied and imaginative responses to the Fermi paradox have been.

No overview of astrobiology would be complete without mentioning the Search for Extraterrestrial Intelligence Institute (SETI). SETI emerged out of a NASA-funded project to search for extraterrestrial intelligences. Out of this project emerged the idea to form a non-profit research organization focused on research and education around the factors of the Drake equation and the search for extraterrestrial intelligences. The SETI Institute was incorporated in 1984 and began operations in 1985. Amongst its first trustees were the previously mentioned Frank Drake (of the Drake equation) and the American astronomer and science educator Carl Sagan (1934–1996) (SETI, 2024). One of the methods the SETI Institute uses to search for alien life involves the use of ground and space-based radio telescopes to scan distant stars for signs of alien life. This nicely brings us onto the 'Wow! signal', the final topic we will discuss in this section on astrobiology.

On 15 August 1977, Ohio State University's 'Big Ear' telescope in the USA was scanning the sky in the direction of the constellation of Sagittarius. A few days later, the astronomer Jerry R. Ehman was reviewing the recorded data when he spotted an anomaly so surprising that he wrote 'Wow!' in the margin next to it (hence the name of the signal). The telescope had picked up a 72-second-long radio signal, with many of the expected hallmarks of a signal having an intelligent extraterrestrial origin. In particular, the Wow! signal was intense or 'loud' (it was 30 standard deviations above background noise at its peak). Its 'shape' was also consistent. It started 'quietly', built to a 'noisy' crescendo, and then gradually diminished, all in a consistent manner. The Wow! signal was also 'sharp', that is, it was a narrowband transmission with a bandwidth of less than 10 kHz. Finally, its frequency was close to the 'hydrogen line'. The hydrogen line is the electromagnetic wavelength of hydrogen. Given hydrogen is the most common element in the Universe, it has often been theorized that an intelligent extraterrestrial species would transmit at this frequency because any technologically advanced species is likely to be familiar with it and thus would recognize it (Kraus, 1979). To date, no uncontentious natural or man-made explanation for the Wow! signal has been identified (Shostak, 2017).

One may wonder on the basis of these considerations why scientists haven't simply concluded that we have received a message from intelligent extraterrestrials. However, there are a number of considerations that might make us doubt an intelligent extraterrestrial origin of the Wow! signal. One of these is that whilst no uncontentious natural or man-made explanation for the Wow! signal has yet been identified, we can't rule these possibilities out. Indeed, at least one of the astronomers at the Big Ear telescope who worked on the Wow! signal considered a man-made origin to be the most likely (Kraus, 1979). Another of these is that, despite our best efforts, we have never been able to find another Wow! signal or any further extraterrestrial 'messages'. It would be rather odd for an extraterrestrial species wishing to make contact with us to simply send one 72-second message only once in 1977.

As a result, the Wow! signal remains frustratingly ambiguous. As it stands, we cannot conclude that it was a message from intelligent extraterrestrials, but nor can we rule this possibility out!

This brings our overview of astrobiology to a close. Hopefully, what has been written here is sufficiently detailed so as to give the reader an overview of the huge variety of different sorts of research that make up astrobiology without being excessively technical. The important thing to note for now is that astrobiological research, as we have seen, ranges from microscopic chemical experiments in a lab (such as the Miller–Urey experiment), to scanning distant stars with enormous telescopes (such as the Big Ear telescope), to using complicated mathematical equations to make various predictions (such as the Drake equation), and much more besides. This is part of what makes astrobiology so fascinating! No doubt much more could be said about astrobiology, but this volume is on exophilosophy, and so it is to exophilosophy we now turn.

This Volume

The rest of this volume is divided into four sections, each focused on a different aspect of philosophy, with each chapter building on the ones before it whilst paving the way for the chapters that come after. The first section (Identity and Identification) opens with a discussion of whether or not life is an observable. Metcalfe and Haramia answer in the negative before going on to consider the implications of this for scientific realism, constructive empiricism, astrobiology, and the life sciences more generally. Chapter 2 by Miller, examines the mysterious topics of Unidentified Flying Objects (UFOs). The chapter begins with an epistemological analysis of Carl Sagan's Extraordinary Claims Require Extraordinary Evidence (ECREE) principle before applying this to the UFO phenomenon. Along the way, the reader is given a detailed overview of the history of UFOs, what we now know, and what we still don't know about this phenomenon. The chapter concludes by pointing out that there are many different ways of thinking about UFOs, far more than either there must be some mundane natural explanation (weather balloons, hallucinations, ball lightning, and so on) or it must be extraterrestrial visitors, and that the options available to us, when thinking about them, will vary depending upon our background philosophical assumptions.

The second section (Ethics and Values) begins with a discussion, in Chapter 3, of whether we would have moral obligations to intelligent extraterrestrials and, if so, what those obligations will be. It argues that we will have moral obligations to intelligent extraterrestrials for the very same reasons we have obligations to one another. In Chapter 4, Howeth builds on the previous chapter by arguing, through a Hegelian lens, that our obligations extend not only to intelligent extraterrestrials but also to all life in the Universe, whilst also tentatively examining what this might look like in practice and the challenges we might face. Chapter 5 examines one particular medium of value, namely music, whilst also preparing the ground for the next section. In this chapter, Stacey begins by considering different philosophical accounts of music. With these accounts in mind, he then goes on to examine the likelihood that an alien would be able to appreciate human music and whether music could therefore provide us with a way to communicate with them. Perhaps disappointingly, Stacey argues for a pessimistic conclusion which somewhat sets the tone for the next section.

Chapter 6 opens the third section (Communication). It argues that meaningful communication using radio signals (or any other similar technologies) over the vast distances of space is likely impossible due to the difficulties of creating a shared language and thus of 'decoding' their messages. Chapter 7 continues in this pessimistic vein by arguing, on Wittgensteinian grounds, that the problem is even worse, and that even if we were to meet an extraterrestrial face to face, meaningful communication may still be impossible.

The fourth section (Religion) examines extraterrestrial life from a different perspective, namely that of religious faith. In Chapter 8, Long demonstrates that the Dharmic traditions (such as Hinduism, Buddhism, and Jainism) are remarkably comfortable with the idea of alien life and even with the idea of it visiting Earth! Chapter 9 examines how the Abrahamic faiths can accommodate the existence of aliens. It argues that whilst these traditions may not embrace alien life with quite the enthusiasm of the Dharmic traditions, they need not be threatened by the discovery of alien life. Finally, in Chapter 10, Bullivant points out that the Ancient Astronaut Hypothesis (AAH) can be seen as a curious species of scripturally literalist, Hickian, technological Euhemerism, with a complicated relationship to atheism, in which, in certain respects at least, aliens have taken the place of God.

A (Cynical) Objection

I want to end this introduction by considering a potential objection a critic might raise against the philosophizing found in this volume. A critic (particularly one in a cynical mood) might query the 'point' or 'use' of all this philosophizing. They might point out that science is clearly valuable because it allows us to develop new technologies which enrich our lives and make them easier (e.g., we owe GPS and modern telecommunications to physicists and astronomers). Similarly, science fiction, in whatever medium, is also valuable because it is entertaining and entertainment is obviously a good. However, so this critic might continue, philosophy, particularly this sort of philosophy, does neither of these things and thus is of limited value and not worth the effort!

Hopefully, given you've read this far, you are not impressed by this objection. Nevertheless, responding to it thoughtfully does help us to clarify the value of exophilosophy and so I will offer a number of points by way of response. The first thing I would point out to such a critic is that philosophy can be, and should be, in at least some sense of the word, fun! In using the word 'fun' I don't wish to trivialize philosophy nor to suggest that it doesn't engage with serious topics in a rigorous manner. Rather, I'm pointing out that many of us love philosophy because we find it endlessly fascinating and thus entertaining. Further, many of us can point to specific moments when we were first introduced to philosophy or to a particular philosophical idea, perhaps when we were young, and simply found it fun! (For what it is worth, for me it was being introduced to the idea of Descartes' malignant demon thought experiment as a teenager! It struck me as an absurd thing to believe, but why is it absurd? And how can we know it isn't the case? There's the rub!) Thus, one response we could make to our critic would simply be to point out that for us at least philosophy is entertaining, and that this is a perfectly respectable reason to pursue it. (And if they are going to continue to be so negative, could they please *leave* so we can continue to philosophize in peace!) If science fiction has value because it is a form of entertainment to science fiction fans, then philosophy has value because it is a form of entertainment for philosophers! At the same time, most philosophers would like to think that we are doing more than simply entertaining each other, and most of us believe our work has some additional intrinsic value beyond mere entertainment. Thus, whilst this response goes some way towards responding to our critique, more is needed.

By way of a second response, I would point out that the criticism our critic has put forward is itself philosophical in nature since it doesn't appear to be a scientific criticism and because it involves appeals to philosophical concepts such as value. Therefore, if this criticism has any value, then philosophy must too, and exophilosophy by way of extension.

However, our critic might respond to this in turn by modifying their claim and making it more modest. They might argue that certain elements of philosophy have practical value, such as ethics or political philosophy, but that, particularly given we haven't yet encountered alien life (and given it may not even exist!), exophilosophy amounts to little more than starry-eyed (literally!) navel-gazing.

I would respond to this by pointing out that this same objection could be levelled against any number of different areas of study in science. One example, tangentially relevant to the topic of this volume, is origins of life research. A critic of origins of life research could point out that it doesn't matter much how life arose; what matters is that life is now here and so we should dedicate our scientific efforts to working out how to improve it in the here and now. Alternatively, directly relevant to the topic of this volume, given it remains unfeasible to ever visit it (at least for the foreseeable future) due to the vastness of the distances involved, why are we putting so much effort into searching for signs of life on distant planets light years away when there are so many much more pressing problems here on Earth.

The problems with such a view are obvious. Origins of life research and astrobiology are immensely valuable for two reasons. First, the questions they try to answer are worth asking and answering (or at least trying to answer) in and of themselves. It is good to try to understand the Universe and our place in it, and origins of life research and astrobiology are attempts to do this. (If our critic refuses to accept even this, then I can only ask them what they think is valuable and why they think we should bother doing anything at all!) Second, scientific knowledge, like all human knowledge, forms an interconnected web with insights from one field of enquiry shedding light onto others, and so on. Even if we never work out for certain how life arose on the early Earth, the insights we gain in trying to work it out will help us understand other questions, which, in turn, may yield new technologies that enrich our lives, and so on.

Exophilosophy is valuable for the exact same reason. In and of itself, attempts to consider the philosophical implications of alien life have value. In addition to this, insights from exophilosophy have implications for other areas of philosophy, as well as biology, linguistics, politics, and countless other important fields of enquiry, to name just a few. The chapters in this volume illustrate this point well. As you shall see, the ideas expressed here have implications for scientific practice, the way we relate to each other and to other living creatures, as well as what it means to communicate and to share a language. In addition to this, they will help to refine our understanding of ourselves and our place in the Universe whilst helping us to answer other questions of (literal) cosmic significance. If that doesn't count as a good reason to study exophilosophy, I don't know what is!

Notes

1 For example, aimed at an academic audience, there is the recently published *Astrophilosophy, Exotheology and Cosmic Religion* (2024) edited by Davis and Faber. There is also Davis's *Metaphysics of Exo-Life* (2023). Aimed at a more popular audience, there is Blackwell's *Alien*

and *Philosophy: I Infest, Therefore I Am* (2017) edited by Ewing, Decker, and Irwin, based on the *Alien* science fiction horror film franchise.

2 The word 'astrobiology' was likely coined in 1935 in an article published in a French popular science magazine by Ary J. Sternfield, a Russian Jew and pioneer of astronautics (Briot, 2012).

3 This also illustrates how origins of life research and astrobiology are closely related disciplines, with many scientists working in both subdisciplines simultaneously.

4 Famously, Urey, the more senior of the two men, insisted that Miller take sole credit for the work out of fear that if they were listed as co-authors, Urey's reputation would result in Miller not receiving due credit for his role in designing and conducting the experiment.

5 The paradox is named after Fermi because, supposedly, in 1950 Fermi and some colleagues at the Los Alamos Lab in New Mexico were having a casual conversation over lunch. Eventually, the conversation turned to UFOs and the possibility of faster-than-light travel, causing Fermi to blurt out, 'But where is everyone?' (or words to this effect since accounts differ and the precise quote is uncertain) (Chace, 2023).

6 In order to appease any pedants amongst our readers, it is worth briefly acknowledging that the Fermi paradox is a paradox only in the more colloquial sense of the word 'paradox' (as opposed to the stricter logical definition of the word). There isn't a *logical* contradiction between the claim that we should expect the Universe to house multiple forms of intelligent life and the observation that we haven't yet discovered or heard from any such intelligences. Nevertheless, such a state of affairs does seem odd and it cries out for an explanation. It is in this looser sense of the word that the Fermi paradox is a paradox.

References

Baxter, S. (2001) 'The Planetarium Hypothesis – A Resolution of the Fermi Paradox', *Journal of the British Interplanetary Society*, Vol. 54, pp. 210–216.

Briot, Danielle (2012) 'A Possible First Use of the Word Astrobiology?', *Astrobiology*, Vol. 12, No. 12, pp. 1154–1156.

Cai, X., Jiang, J. H., Fahy, K. A., and Yung, Y. L. (2021) 'A Statistical Estimation of the Occurrence of Extraterrestrial Intelligence in the Milky Way Galaxy', *Galaxies*, Vol. 9, No. 5, pp. 1–14.

Chace, Calum (2023) 'The Fermi Paradox: Where Is Everyone?', *Forbes*. Available at: forbes.com/sites/calumchace/2023/01/04/the-fermi-paradox-where-is-everyone/ (Accessed 27/05/2024).

Davis, Andrew M. (2023) *Metaphysics of Exo-Life: Toward a Constructive Whiteheadian Cosmotheology*, Grasmere: SacraSage.

Davis, Andrew M. and Faber, Roland (Eds.) (2024) *Astrophilosophy, Exotheology and Cosmic Religion: Extraterrestrial Life in a Process Universe*, Lanham: Lexington Books.

Ewing, Jeffrey and Decker, Kevin S. (2017) *Alien and Philosophy: I Infest, Therefore I Am*, Hoboken: Wiley-Blackwell.

Forgan, Duncan H. (2017) 'The Galactic Club or Galactic Cliques? Exploring the Limits of Interstellar Hegemony and the Zoo Hypothesis', *International Journal of Astrobiology*, Vol. 16, No. 4, pp. 349–354.

Kaufman, Marc (2022) 'About Astrobiology: Life, Here and Beyond', *Astrobiology at NASA: Life in the Universe*. Available at: astrobiology.nasa.gov/about/ (Accessed 24/05/2024).

Kraus, John (1979) 'We Wait and Wonder', *Cosmic Search*, Vol. 1, No. 3, pp. 31–34. Available at: bigear.org/CSMO/PDF/CS03/cs03p31.pdf (Accessed 03/06/2024).

Miller, Stanley L. (1953) 'A Production of Amino Acids Under Possible Primitive Earth Conditions', *Science*, Vol. 117, No. 3046, pp. 528–529.

NASA (2024) 'FAQ: How Has the Astrobiology Program Evolved over the Years?' *Astrobiology at NASA: Life in the Universe*. Available at: astrobiology.nasa.gov/about/faq/astrobiology-program-evolved/ (Accessed 24/05/2024).

SETI Institute (2024) 'History of the SETI Institute', *SETI Institute*. Available at: seti.org/history-seti-institute (Accessed 02/06/2024).

Shostak, Seth (2017) 'Was It ET on the Line? Or Just a Comet?' *SETI Institute*. Available at: seti.org/was-it-et-line-or-just-comet (Accessed 03/06/2024).

Shostak, Seth (2021) 'Drake Equation', *SETI Institute*. Available at: seti.org/drake-equation-index (Accessed 24/05/2024).

Soffen G. A. (1997) 'Astrobiology from Exobiology: Viking and the Current Mars Probes', *Acta Astronautica*, Vol. 41, No. (4–10), pp. 609–611.

Ward, Peter and Brownlee, Donald E. (2000) *Rare Earth: Why Complex Life Is Uncommon in the Universe*, New York: Copernicus.

Yu, Chao. (2015) 'The Dark Forest Rule: One Solution to the Fermi Paradox', *Journal of the British Interplanetary Society*, Vol. 68, pp. 142–144.

PART I
Identity and Identification

1

IS LIFE OBSERVABLE?

Thomas Metcalf and Chelsea Haramia

Introduction

As we move through the world, we perceive various medium-sized objects, such as cats, staplers, airplanes, and donuts. Our visual acuity does not permit us to discern individual electrons nor virions. Thus, there seem to be scale-based limits to what we can perceive in our everyday awareness. Beyond these examples, various phenomena can be approximately measured and tracked using our own mental assessments or measurement devices, but do not present themselves as clearly discernible visual or auditory phenomena. For example, time's passage itself does not seem to sound like anything, but most of us can hear our phone's alarm clock when it beeps. Arguably, ordinary space does not look like anything, although a meter stick is visually perceptible. Broadening our scope to entities that are not fully within the purview of natural science, we might hold that causation, laws of nature, and the persistence of persons over time are similarly invisible to our senses.[1]

In this chapter, we argue that life is unobservable as well. In the same way that scientists can infer that time has passed or that an electron is present in a cloud chamber, they can infer that some collection of particles is alive. But they cannot *perceive* unaided whether life is present in any location or object. Thus, life is unobservable. The principal implication for astrobiology is that, if we wish to use science to detect life, we need to understand whether and how to use science to detect unobservables.

Someone might wonder at this point whether anyone thinks life is *observable*. Are we attacking a straw position? Well, it's likely that many or most of us *don't* yet think that life is *unobservable*. The question does not appear to have been seriously addressed in the literature up until now. Therefore, we will first establish that life is, in fact, unobservable, which will also help us understand the phenomenon of observability more generally. After that, we will explain which further conclusions we should draw about philosophy, science in general, and astrobiology in particular.

DOI: 10.4324/9781003440130-3

The Observable–Unobservable Distinction

It is rare to encounter a scientist who denies that electrons, gravity, or viruses exist. Most ordinary people would also grant that these entities exist. If we take the commonsense position that electrons, gravity, and viruses exist, but cannot be directly observed in the intuitive sense in which we observe medium-sized physical objects, then we believe in unobservables. This position appears to be the strong majority position among philosophers in general and an even stronger majority among philosophers of physical science in particular.[2] Therefore, for the purposes of this chapter, we feel safe in taking seriously the possibility that unobservables exist, although we will need to say more about whether there is a useful distinction to be made here.[3]

The bulk of the debate in the philosophy of science about observables and unobservables concerns whether science *achieves* knowledge of unobservables; whether it is the *goal* of science to produce knowledge of unobservables; and whether certain views in the philosophy of science are *compatible with* relying on that distinction (Chakravartty, 2023; Monton and Mohler, 2023; van Fraassen, 1980). Many critics question whether constructive empiricists—those who hold that the goal of science is to achieve knowledge of observables alone (van Fraassen, 1980, pp. 8–12)—can coherently appeal to the distinction between observables and unobservables (Musgrave, 1985; Muller, 2004; Dicken and Lipton, 2006). We are not writing from the perspective of constructive empiricism, so we can set many of those objections aside. However, our thesis is that life is unobservable, and we do believe that some objects *are* observable, so we cannot avoid entering the debate over whether there is a coherent distinction there overall.

The observable–unobservable distinction is commonly understood in terms of the unaided–aided distinction in perception (Chakravartty, 2023, § 1.1; Monton and Mohler, 2023, § 1.6; van Fraassen, 1980, p. 15). Detection as aided by a microscope or cloud chamber does not count, in this sense, as observation. However, we believe there are other ways in which some phenomenon can count as unobservable. Life, unlike electrons and viruses, seems to belong to a different category of unobservables than the ones in which unobservability depends on whether our perception is aided by some instrument. Let's consider four types of potentially existent entity as shown in Table 1.1.

We expect most readers to regard only the items in Group 1 to be uncontroversially observable.

Whether the items in Group 2 are observable is debatable. Maybe a human could undergo an eye-and-brain operation to allow them to perceive radar waves (cf. Churchland, 1985, pp. 44–45). Maybe an extraterrestrial species evolved to be able to perceive more of

TABLE 1.1 Examples of Unobservable Entities

Group 1	*Group 2*	*Group 3*	*Group 4*
• Airplanes	• Electrons	• Phases of matter (for example, gasses)	• Axiological badness
• Cats	• Gravity	• Intelligence	• Merely possible worlds
• Donuts	• Radar waves	• Religions	• Platonic numbers[a]
• Staplers	• Space	• Types of material (for example, metals)	• Platonic properties
	• Time		
	• Virions		

[a] van Fraassen (1980, p. 15) regards this as a paradigm unobservable. For clarity, by "platonic," we mean to say that the numbers are abstract or non-spatiotemporal.

the electromagnetic spectrum unaided. And when we perceive virions with the use of an electron microscope, the electrons are performing a role analogous to the role of photons when we observe medium-sized objects. We can generally feel acceleration, but we cannot distinguish it from gravity (or, if gravity is a type of acceleration, then we cannot distinguish non-gravity acceleration from gravity). One might argue that proprioception is perception of space, and that humans have some ability to gauge how much time has passed, but others might reply (in a Kantian (1998 [1781], A26, A33, A239) vein)[4] that we can only conclude that time and space exist because time and space are preconditions of any perception at all. In turn, we perceive objects and events that have some order, and we might conclude on that basis that space exists or that time has passed. Thus, some people will question whether everything in Group 2 is unobservable, and others will object to the observable–unobservable distinction based on items like the ones in Group 2.

In stark contrast to the items in Groups 1 and 2, the items in Group 4, if they exist at all, are arguably impossible to observe empirically, no matter what instrument you might build.[5] However, they are only rarely mentioned substantively in scientific practice. Even numbers themselves are rarely referred to directly in scientific practice; instead, written numerals are used to represent quantities or magnitudes, and the quantities or magnitudes are taken to be explanantia or explananda. Possibly, mathematicians take themselves to be discovering interesting properties of particular numbers, but mathematicians' methods are generally not empirical. Thus, there is little reason in the context of this chapter to work hard to decide whether Group 4 items are observable in any interesting sense.

We believe that life, as an unobservable, most naturally fits into Group 3. It is a central concept in scientific practice—indeed, an entire top-level field of natural science is devoted to it—but our inability to perceive whether a collection of proteins is *alive* is not merely the result of our not being able to build a sensitive-enough detector. Analogously, most social scientists grant that religions exist, but they would not say that there is a kind of *religion detector* that lights up in the presence of religious people or moves a needle when one travels to more-religious regions of the world. And in the search for extraterrestrial intelligence (SETI), SETI practitioners do not aim to detect intelligence per se; they are searching for technology and using technology as a proxy for intelligence. As astronomer and SETI founder Jill Tarter notes, "SETI is not the search for extraterrestrial intelligence. We can't define intelligence, and we sure as hell don't know how to detect it remotely" (Cofield, 2018).

In Group 3, there are no borderline cases of perceptibility, such as those that depend on the distinction between perceiving with the use of a light bulb versus a microscope. Similarly, in the cases of phases of matter or materials, what we do is perceive various bottom-level physical properties or features and infer on that basis that the material in question meets our definition of "metal" or "gas." But the metalness of stainless steel does not somehow *look* like something *else* over and above its shine, no matter how you aid your perception.

We might, therefore, append some further descriptions to the groups we surveyed, as follows:

Again, we assume that the items in Group 1 are observable. The items in Group 2 are unobservable unaided, but we use instruments to observe them, and some creatures may be able to observe them unaided. The items in Group 3 can be called "constitution-based unobservables," because while we observe the parts of the objects (and, on that basis,

TABLE 1.2 Categories of Unobservable Entities

Group 1: Observables	Group 2: Instrument-Based Unobservables	Group 3: Constitution-Based Unobservables	Group 4: Nonphysical Unobservables
• Airplanes • Cats • Donuts • Staplers	• Electrons • Gravity • Radar waves • Space • Time • Virions	• Phases of matter (for example, gasses) • Intelligence • Religions • Types of material (for example, metals)	• Axiological badness • Merely possible worlds • Platonic numbers • Platonic properties
Currently observable	Observable with different sensory apparatus	Unobservable even with different sensory apparatus, but important to science	Unobservable even with different sensory apparatus, but peripheral to science

conclude that they compose the phenomenon in question), those parts could generally compose items that aren't examples of the phenomenon, and the phenomenon itself is imperceptible. The items in Group 4 are commonly regarded as nonphysical or otherwise not amenable to any empirical study, even in principle (see Table 1.2).

While it may be difficult to draw principled distinctions between perceiving the items in Group 1 versus Group 2, and such a distinction may depend on a questionable distinction between aided and unaided perception, we do not see how it could be difficult to draw a principled distinction between perceiving the items in Group 1 and perceiving the items in Group 3. Our perceptions or detections of those Group 3 items do not depend on simply employing a device that works in any way analogous to our familiar senses such as sight and hearing. There is no obvious example of some everyday device, such as a light bulb, that permits our perception of something like intelligence in a way that allows us to count it as perceptible with the aid of some devices but not with others. Instead, we could administer some kind of intelligence test and perceive the results of the test, typically displayed on a computer monitor or piece of paper. But we're not empirically perceiving the intelligence itself, even if we are perceiving the computer monitor that displays the results of the test.

One last way to draw distinctions between the groups is to notice that Group 3 comprises items that are not *perceptually* borderline cases as in Group 2, but they are *definitionally* borderline. As for Group 1, generally, we do not argue about whether something is or isn't a stapler, but we might argue about whether something (for example, a hot dog) is or isn't a sandwich, even if sandwiches are straightforwardly observable, and (given some definition) we can straightforwardly observe *that* something is a sandwich.[6] Thus, there seem to be definitionally borderline cases therein. The items in Group 2, in contrast, seem to generally be scientific terms of art, which means there is less debate about whether some entity counts as a member of the category. As for Group 3, people certainly disagree about whether something is a religion, and whether something is a metal.[7] In some of these cases, we might decide that there is no single, correct answer. And last, as for Group 4, there seems to be no real debate over whether some object counts as a member of that kind. To speculate, the reason might be that we don't really *encounter* the items in Group 2 and Group 4 and *then* have to decide what we think about them. Instead, we tend to infer their existence, or their existence appears to us via reason or intuition, or the entity must exist in order to explain some observable phenomenon.[8] And we conclude that, because of the role it plays or the phenomenon it must explain, the entity must be a radar wave, a gravitational force, or a merely possible world.

Before we continue, we should make one more note of clarification. When we say that something is unobservable, we don't mean that it is impossible to observe anything that *entails* its presence. For example, we agreed earlier that cats are observable. But cats are alive—therefore, isn't life observable? In reply, we do not believe that observability is transmissible in that way; indeed, it would be something like the Fallacy of Division to hold that because some whole is observable, all of its properties are observable. Otherwise, we might have to say that because cats are observable, and elements in cats' bodies emit beta particles, beta particles are observable. Someone might counter-reply here that whether cats are alive is an *analytic* or conceptual question, unlike the *synthetic* fact that their bodies emit beta particles. But as before, that doesn't seem sufficient to render life itself observable. For one thing, a small child might have a conception of "cat" without having any conception of the distinction between alive and not-alive, so it's not clear to us that cats' being alive is an analytic truth. And for another, intuitively, someone might observe an object that analytically has a certain feature, without the feature's being observable. I might observe a chiliagon, but arguably, its having 1000 sides isn't observable, at least not to humans. I might observe a bacterial colony in a petri dish, and it might be that bacterial colonies analytically comprise bacteria, but the bacteria aren't observable.

We will argue that life is unobservable in general, and, in particular, much more like the items in Group 3 than the items in Group 2 (and in Group 1). Given that life belongs in Group 3, there will also be no interesting question of whether it might be a borderline case between observables and unobservables. We turn, then, to the questions of whether life truly is unobservable, and if so, in what way.

Why Life Is Unobservable

We are arguing that life is unobservable. We have three arguments for that thesis: the argument from empirical unobservability, the argument from resemblance, and the argument from dispensability.

The Argument from Empirical Unobservability

One reason to say that life is unobservable is that there is no biological sensory apparatus, nor scientific instrument, that could, even in principle, physically interact with the "life" in a living organism. Like the other constitution-based unobservables, we can observe the parts, and we commonly believe that certain quantities or configurations of the parts are evidence of life, but nothing detects the life per se. In principle, there could be such a thing as élan vital (cf. Normandin and Wolfe, 2013), or a kind of substance or force that all and only living things possess, but scientists have no good evidence for it. So, we aren't aware of any substance that a scientific or perception-based instrument could detect.

To begin with biological sensory apparati, there is no particular color or shape that living things have and nonliving things don't. We cannot name a certain frequency, such as 440 Hz, that living things emit and nonliving things don't. And for any candidate example, we can imagine a nonliving thing that nevertheless emitted that phenomenon. Extremely realistic robots could look just like humans; they would look exactly like living things to unaided perception, and there would be no extra perceptual stimulus associated with *life* that was missing.

Constructive empiricists generally regard entities that are only detectable by scientific instruments to be unobservable (Monton and Mohler, 2023, § 1.6). But even if we consider scientific instruments to permit a kind of observation, we believe we should reach the same conclusion about life. Our instruments can report the presence of certain chemical elements, but many nonliving things are composed of the same chemical elements as uncontroversially living things. A comet might contain carbon, hydrogen, oxygen, and nitrogen, but certainly not be alive. The same is true, *a fortiori*, for the presence of protons, neutrons, and electrons. One might argue that certain types of chemicals, such as amino acids, are hallmarks of life, but of course it is possible to synthesize tanks of amino acids without producing anything that we would call "alive." One might even arrange these proteins in structures that resemble living organisms, but that's not sufficient, because a corpse is arranged in essentially the same structure as a living organism. Such a corpse could be constructed, like a Frankenstein's monster, in a way that it was never a living organism. Someone could argue that at least there are observable structures, such as cells, that entail the presence of life. (Some cells, such as chickens' eggs, are observable to the unaided eye.) But we wouldn't want to say that an unfertilized chicken's egg is alive. And entities could contain cells without being alive; for example, a petri dish may contain millions of bacterial cells, but the petri dish itself is not alive. And we certainly wouldn't say that an entity must be composed entirely of cells in order to be alive; a human with a pacemaker or dentures is still a living organism. Indeed, some people would be inclined to describe some entities with *no* cells at all as "alive." If we think that a living being's "survival" from time t1 to t2 means that a being is alive at t1 and alive at t2, then the classic philosophical question of whether it is possible to survive the death of one's physical body is partly the question of whether a being with no cells at all— such as a being that was a disembodied spirit or uploaded onto a computer—might be alive.[9]

One might therefore, at least, attempt to define life in terms of its activities or processes rather than its composition (Mix, 2015). But we can build devices that convert carbon dioxide molecules to oxygen molecules. We can build self-replicating entities, such as simple computer viruses (even those that experience natural selection), that most of us would not consider "alive." Toy dolls that excrete are available. For these reasons, we believe, the presence of life must remain, like the other items in Group 3, something that is fallibly inferred from a collection of other activities. And we take seriously the powerful criticisms of the very project of coming up with a strict or universal definition of "life" (Cleland, 2019, ch. 5).

Someone might object to this argument on the following grounds:

[Objection] But if life objectively exists, then there are objectively existing physical properties and activities that entail that something is alive. At a certain level, it must be possible to build a detector that is sensitive and specific enough to successfully register "Life" when life is present and "Nonlife" when it isn't.

We do not believe we need to take a strong position on whether life *objectively* exists, but we believe that it does—that is, that there is a single correct answer to the question, for example, of whether there are any living things on Earth. Therefore, let L be the set of substances and activities that $n\%$ sensitively and specifically distinguish living from nonliving organisms, where "$n\%$" is whatever level we regard as necessary and sufficient for a "successful" detector, and a successful detector is one that allows us to observe something with fewer than 5% false positives and fewer than 5% false negatives. And suppose that a "successful" detector is at least $m\%$ likely to light up in the presence of L and no more than $1 - m\%$ likely to light up when L is not present, where $m\%$ is whatever our threshold for "success" should be. Perhaps it is possible to build a successful L-detector in that sense.

We reply that the possibility of building such a device is not normally taken to show that some phenomenon is *observable*. Recall, it's standard to distinguish observables from unobservables in terms of whether they are perceptible unaided. Still, this might invite a further reply based on some of the dialectic outlined earlier: Perhaps some human could be trained to detect L with $m\%$ accuracy.[10] We respond, in turn, that this is still not normally enough to say that some phenomenon is *observable*. Maybe an expert sociologist can detect with 95% accuracy whether a group counts as a religion. That doesn't mean that anyone is empirically observing the religion-ness itself; indeed, the sociologist might only be reading about the group in some book. By analogy, a human could be trained with over 95% accuracy to decide whether there is a prime number from 1 to 20 of coins on a table, but that human is not empirically perceiving the *primeness* of the number. An unusual human could infallibly form true beliefs about where the president is at any given time, but would not necessarily be empirically observing the president's location.[11] Sensitive and specific *detection* is not sufficient for empirical *observation*.

The Argument from Resemblance

Our second argument is that life bears a very strong resemblance, in key ways, to other unobservables. In particular, it is very similar to certain natural kinds or types, such as gasses and metals, that are identified by their physical properties and behaviors, but not directly detected by any scientific instrument. It also bears a strong resemblance to the constitution-based unobservables, such as intelligence and religions, that are perhaps not natural kinds but still detected based on constitution.[12] As noted earlier, philosophers and scientists debate about the correct understanding of the phenomena, such as religions or intelligence, in Group 3. Citing the phenomenon can help with higher-level descriptions in the field in question, but the concept does not appear to be vital to scientific investigation "on the ground." And one might reasonably wonder whether there is any objective definition of the phenomenon, and whether there are too many borderline cases and counterexamples in order to make the concept useful (Cleland and Wilson, 2013, p. 21). If gasses, metals, intelligence, and religions are unobservables—and we believe it's plausible that they are— then that's further evidence that life is unobservable, because it shares a lot in common with those concepts.

The Argument from Dispensability

Third, it's worth stepping back for a moment and remembering what work the observable– unobservable distinction is supposed to do. Recall that constructive empiricists, in particular, want to be able to argue that science pursues knowledge of observables but not of unobservables. In turn, we should expect that observables are vital to scientific practice, while unobservables are relatively dispensable or reducible, in the sense that scientific practice can proceed more or less as normal while being agnostic about the existence of the unobservables. Chemists can describe the outcomes of chemical reactions without committing to the real existence of electrons, and physicists can calculate the positions of planets without claiming to see gravity per se. However, it would be difficult to perform standard chemistry without committing to the existence of thermometers or litmus paper, and it would be difficult to perform standard astronomy without committing to the existence of planets.

We believe that life, in this sense, looks more dispensable or reducible than indispensable or irreducible. Even in biology, scientists can talk in terms of individual organisms, their health markers, their diets, their fecundity, and so on, without ever saying that these organisms are *alive*. We do not commonly find, in research articles, tables that count "alive" as 0 and "nonalive" as 1 for the purposes of regression analysis. Nor does "liveness" ever appear as a numerical variable such that we can compare a mean in some population to a mean in another population and decide whether there is a statistically significant difference. Scientists who are interested in extraterrestrial life attempt to find such life by looking for biosignatures and technosignatures, not by pointing a telescope built to detect "life waves" at a distant planet. Thus, this is a third respect in which life seems a lot like other unobservables. Of course, we will return later to the question of whether we actually *should* try to remove talk of "life" from our sciences; at this point, we simply want to say that life resembles other unobservables in this respect as well.

We believe that taken together, these considerations make it very difficult to view life as observable, so we should proceed to consider what kind of unobservable it is.

How Life Is Unobservable

Recall our distinction between types of potentially observable entities. The items in Group 1, such as airplanes, are uncontroversially (at least, for our purposes here)[13] observable. The items in Group 4, such as platonic numbers, are uncontroversially (again, at least, for our purposes here)[14] unobservable, but only peripherally relevant to our common scientific pursuits. The items in Group 2 are observable with the use of scientific instruments that are closely analogous to everyday objects such as light bulbs and somewhat analogous to our standard sensory modalities, but cannot be observed unaided. The items in Group 3—the group that, we argue, contains life—are not clearly directly observable, not even with the use of scientific instruments that approximate (but modify or intensify the acuity of) our natural sensory apparatus.

There is another reason, which we discussed in the previous section, to think that life belongs in Group 3. It's something like reducibility to lower-level features. Whether a collection of particles counts as alive seems to be a matter of whether its lower-level physical features are sufficient to lead us to count it as alive. In that respect, it appears very similar to the categories of "gas" and "metal." When we inspect candidate definitions of "life" in the literature, we find the same phenomenon. There is a set of potentially observable lower-level features—such as metabolism and reproduction—that are sometimes taken to be strong evidence of life (Joyce, 1994; Mix, 2015, § 4; cf. Cleland, 2019, ch. 1; Mariscal, 2023, § 1.2). Similarly, it's common to use IQ tests to try to estimate people's intelligence (Whitten, 2020), even though intelligence is more than IQ alone (Bond, 2009).[15] It is common to identify religions based on the presence of family-resemblance features,[16] which presumably would be unnecessary if religion were directly observable.

Yet another reason to place life in Group 3 is that it resembles the other entities in another respect. There are questionable or borderline cases of gasses and metals, and there are questionable or borderline cases of life. For example, different fields of science consider nitrogen to be or not to be a "metal" (Swinburne University of Technology, n.d.), and chemists regard a set of elements as "metalloids."[17] Similarly, viruses and prions are widely regarded as interesting borderline cases about life: they are life-like in many respects, but also very different from paradigm examples. Perhaps artificially intelligent computer programs would also be interesting borderline cases. In this respect, once again, life appears to be like the other constitution-based unobservables. Recall, from Section 2, that the items in Group 3 tend to have definitional borderline cases, not perceptual borderline cases. We are not saying that it is questionable whether we are truly perceiving the phenomenon unaided; we are saying that people may actually be correct to debate about whether the entities count as alive (or as a metal, or a religion).

Someone might argue here that life is only unobservable in the relatively uninteresting ways that *conventional* facts are unobservable.[18] Perhaps it's true by convention that currency is valuable, for example, that some car is worth 15,000 dollars or euro. Similarly, perhaps it's only true by convention that 5 is a prime number, because numbers are our ideas, and we made up rules that entail that 5 is prime. We don't empirically observe that value when we look at the car, and we don't empirically observe the primeness when we look at sets of five objects. In turn, it might only be a convention that we call some objects, but not others, alive.

Our reply, at this point, is that most of what we say in what follows will apply even if whether something is alive is merely a convention. But we also add that life, and what's associated with it, does not seem to behave the way that typical conventions do. If we all decided tomorrow that dollars are worthless, then they would become worthless; but if we all decided tomorrow that birds aren't alive, they would still behave very much like living organisms—indeed, we think, they would simply still *be alive*. We would also say that it seems obvious to us that living things existed ten million years ago, even before social conventions did, and that we're not merely retroactively applying a concept; we think it's obvious that *it was true ten million years ago* that living things existed.

Perhaps in a similar vein, someone might argue that we are really only showing that "life" is *difficult to define*. We reply that none of our main three arguments in the previous section was chiefly about whether "life" is definable. Arguably, many observables, such as chairs and sandwiches, are difficult to define rigorously, while many or most traditional unobservables, such as electrons, do not have particularly mysterious definitions. Granted, the Group 3 unobservables, such as metals and religions, might be difficult to define. But again, that's not in itself what makes them unobservable. Only if someone constructed a definition of "religion" that made it depend fully on observable features might we be tempted to say that religions were observable—but in our view, it is a priori implausible that such a definition would be widely accepted. It's not that it's difficult to define "religion" (although it is); it's that whatever the correct view of religions is, it's a view of a phenomenon that is not empirically detectable, at least not with anything like the perceptions and instruments we have today.

Lastly, someone might also argue that life is observable, but not "observable that", following a distinction made by van Fraassen (1980, p. 15). His analogy is of a tennis ball. If a tennis ball traveled 100,000 years back in time and were encountered by Stone Age humans, then the humans would certainly observe *it*, even though they would not observe *that* it was a tennis ball. Another example comes from Churchland (1985, pp. 36–37). If witches existed, people could certainly observe them, even though no one could observe that someone was a witch. Perhaps life is the same way. We observe living things all the time. Maybe we don't empirically observe *that* they are alive, but could we empirically observe their *living*? We have two replies.[19]

First, if we want to say that tennis balls are observable, even though not everyone will observe that something is a tennis ball, that doesn't tell us that the same will be true about life. Tennis balls are like the examples in Group 1, such as staplers. There is little debate over what counts as a tennis ball, and the properties that make something a tennis ball tend not to depend much on processes over time nor on microscopic properties. In these respects, tennis balls are quite a bit different from plausible attempts to understand or define life (cf. Mariscal, 2023, § 1.2). Still, there is the case of witches—surely witches would be observable, even though what makes them witches is not easily observable. In reply, as with tennis balls, there is nevertheless little debate over what makes something a witch. And if merely having some sort of sensory, causal contact with an entity is sufficient to count as having observed it, then even photons and electrons are observable, which seems wrong. (One could see electrons when one sees an arc of electricity, or see photons when one sees reflected sunlight off a shiny object.) We therefore believe that one needs to be able, to some degree, to observe *that* something is a witch in order for "witches are observable" to be true. Or, perhaps put another way, one needs to be able to observe the person *as* a witch, or the arc *as* electrons.

Second, suppose we were to grant that life is observable, but it is not observable *that* something is alive. It will still be important to realize that whether things are alive is unobservable, largely because we care (and astrobiologists care) about whether we can detect things that are alive. Our overall conclusions about the implications of life's being unobservable will remain essentially unaffected. We turn now to consider those implications.

Why It Matters That Life Is Unobservable

We have argued that life is unobservable. In particular, the presence of life is unobservable in a similar sense to how the presence of a gas or metal per se, or a religion, is unobservable. Possibly, this conclusion is interesting in itself. We can add a new category to our standard list of unobservables, and indeed, a category of special importance to most of us. But we believe that this conclusion has certain, important implications for philosophy, for scientific practice, in general, and for astrobiology, in particular.

Philosophy

The observable–unobservable distinction, or the problem of unobservables, is central to the debate over constructive empiricism. If, for example, there were no unobservables, then most of the motivation for constructive empiricism would disappear. We could cheerfully accept scientific realism. But our adding "life" to the list of unobservables may lend support to, or undermine, a case for constructive empiricism. If we decide it is not the job of science to describe life, then, if this is acceptable, then constructive empiricism will be unscathed. But if this is implausible—if we think, for example, that biology is undeniably a science, and that biology is undeniably about life per se—then this will count against constructive empiricism.

Other areas of philosophy may also be affected by the conclusion that life is unobservable. Life is important in some debates in ethics and in social and political philosophy, as well as environmental philosophy. For example, some of the popular debate over abortion has to do with whether an embryo is alive and whether it is a living organism.[20] Similarly, some theorists in environmental ethics regard life per se as more valuable or important than nonlife.[21] The life–nonlife distinction may also be relevant to some topics in metaphysics, such as constitution and diachronic identity or persistence.[22] And even the philosophy of religion comprises some debates that count life as important versus nonlife.[23]

The effect of regarding life as unobservable may, therefore, be to somewhat problematize some traditional debates in philosophy. However, we do not expect any kind of catastrophe. The reason is that traditional philosophical methods are arguably largely a priori,[24] and so it will not require an important change in our methods. The unobservable nature of life can be accommodated within something like business as usual. Because philosophy is fundamentally already about concepts, hypothetical cases, and intuitions, it comes equipped already to handle such unobservables. Constructive empiricism about science is interesting because it is interesting to consider whether science can proceed without knowledge of unobservables, but relatively few philosophers would care much about the question of whether philosophy can proceed without knowledge of unobservables.[25]

Lastly, whenever we learn that philosophy has a role where we didn't previously acknowledge it had a role, this reminds us of the usefulness of philosophy and its connection with the natural sciences. Thus, this would provide some support for naturalism, very broadly construed in philosophy, suggesting that there is no strict border between the natural sciences and philosophy. In this case, philosophical methods would have a greater role in life-related sciences than we thought they did. But the border is open in both directions. On the one hand, naturalists and experimental philosophers have argued that our biological or biochemical knowledge might require our questioning or revising philosophical conclusions.[26] Yet on the other hand, taking philosophy seriously in life-related sciences might mean rejecting an ostensibly scientific conclusion because it conflicts with philosophical assumptions or hypotheses. (Some of us will be more comfortable with this result than others are.) If we use a hypothesis that life is present on some planet in order to draw some conclusion about what else takes place on that planet, then a largely philosophical argument that we don't know whether life is present after all will require us to modify our scientific conclusion about what else we thought about that planet. We turn now to further exploration of the implications of our conclusion for the sciences.

Science and Life-related Sciences

The effect of explicitly regarding life as unobservable may be different in the sciences from what it is in philosophy.

Suppose, contrary to constructive empiricism, that science actually aims at knowledge of unobservables. How do we acquire such knowledge empirically? In the case of Group 3 (such as gasses, metals, and religions), we typically discover some observable phenomena that we associate with the unobservable phenomenon at issue, and on that basis, conclude that the unobservable phenomenon is present. In turn, on that basis, we can make further predictions, draw further conclusions about the phenomenon itself, or argue for practical implications about it. For example, whether something is a religion may affect whether we think a group of practitioners should be legally required to pay taxes. Often, such questions are resolved, at least partially, by appeals to intuition, common sense, or hypothetical cases (and our intuitions about them) (cf. Pust, 2023, § 2.2). We've already seen examples of this method in previous sections of this chapter, in which (for example) we noted the question of whether an unfertilized chicken's egg is alive. And similarly, in the case of religion, we might draw analogies with other practices or institutions (such as corporations and governments) in order to decide whether some group is actually a religion. To use another example from Group 3, we might appeal to paradigm examples of people or other animals we consider to be intelligent, and then ask the degree to which some candidate creature or person resembles those exemplars.[27] Thus, in order to incorporate, in some scientific research, the datum that some object is (or isn't) alive, we may end up appealing to intuitive judgments about analogies or paradigm exemplars: characteristically philosophical methods.

When a scientific field starts to require more use of reason, intuition, and hypothetical case arguments in order to make progress, it starts to stray more into the territory of philosophical analysis. This can occur because we need to use philosophical methods to analyze definitions, and because we need to use philosophical methods to decide on principles. Philosophers have a long history of using hypothetical cases to decide whether some phenomenon counts as "knowledge" or "survival," and so it would be natural to use similar cases to help decide whether some object is alive (cf. Pust, 2023, § 2). And similarly, philosophers participate in long-standing debates over philosophical principles, such as Ockham's Razor, that are taken to tell us a lot about real-life scientific phenomena.[28] Thus, again, philosophy would play a larger role in the life-related sciences than we may have originally thought.

Some of us would say that this is a welcome development. Many philosophers are familiar with the elementary mistakes made by science popularizers who have little acquaintance with real-life philosophical practice and content (Goldhill, 2016; Shaw, 2016). Thus, we might think that an injection of philosophical theorizing would invigorate and nourish some scientific fields. Another likely effect would be to make traditionally concrete, hands-on scientific fields, such as biology and biochemistry, more like their more abstruse and abstract cousins, such as theoretical physics. After all, arguably philosophical assumptions, such as the Principle of Relativity and Ockham's Razor, are foundational to theorizing about relativity.[29] Theorists about quantum mechanics can say very little about what's actually going on—in terms of the physics that affect all of us—without entering into a chiefly nonempirical debate. They might tell us that there is an unobservable—for example, a pilot wave—that nevertheless determines something that happens in the physical, observable universe.[30] Or, they might argue that the Schrödinger equation, when suitably understood, entails the near-constant production of a set of unobservable (by us), parallel branches of the universe.[31] This fascinating debate in theoretical physics might suggest that parallel debates are necessary or important in biology and biochemistry. Thus, biology and other life-related sciences may start to forge stronger connections with philosophy, or their philosophical commitments may start to be more evident.

Of course, if we are constructive empiricists who want to do science and the life-related sciences, then we have three options. We can argue that life is observable; we can abandon constructive empiricism; or we can decide that science should jettison talk of "life." This last option would probably require a radical revision of our ideas of biology and its subfields or related fields, although someone might reply that constructive empiricism is already revisionary.[32] But arguably, whether something is alive tells us something important. We might lose something if we start merely talking about life-adjacent phenomena. After all, we assume that the general public would be more interested in an announcement that life is found on other planets than an announcement that whatever substrate we consider closely related to life has been found on other planets. Thus, being able to find life, and apply the knowledge that we've found life, is likely to be scientifically productive, and it's difficult to be comfortable abandoning that terminology and associated knowledge.

Someone might argue here, finally, that our conclusions won't change the role of "life" in the sciences much, because "life" was already difficult to define rigorously. But as we explained in Section 3, there is a clear difference between unobservability and indefinability. It's very common to discuss some observable entity (such as chairs and sandwiches) even if, when pressed, it is difficult to give a rigorous definition of it. We would add that there is little discernible debate in the philosophy of science about whether the aim of science is to pursue knowledge of rigorously definable phenomena, but as we've seen, there is substantial debate over whether the limits of science should be drawn at the limits of observability.

Astrobiology

Lastly, we believe that our conclusion is especially important for the field of astrobiology.

Let's return to the debate between scientific realists and constructive empiricists about the aims of science. For the realist, the proper goals of science include knowledge of unobservables; for the constructive empiricist, they include only knowledge of observables. We can imagine a corresponding scientific realist about astrobiology and a constructive empiricist about astrobiology. If astrobiology is fundamentally about an unobservable, then the scientific realist will have to say that astrobiology needs to employ some kind of method of acquiring knowledge that goes beyond empirical observation.[33] Then again, if astrobiology can be conceived of as pursuing knowledge of some observable phenomena, then it is compatible with constructive empiricism to pursue astrobiology.

We wouldn't want to impute our personal views to all astrobiologists, so we need to consider all the possibilities in the logical space. Astrobiology is the search for and study of life in our universe, including terrestrial life. It is commonly hoped, of course, that this discipline is capable of achieving the discovery of extraterrestrial life as well. But the search for extraterrestrial life is not merely the search for life per se. This search also includes the search for and study of evidence of life and life-supporting conditions. It's an open question whether *evidence* of life and life-*supporting* conditions are, themselves, observables. In turn, given our conclusion that life is unobservable, the distinction between scientific realism and constructive empiricism determines four possible combinations of answers to questions of purpose.

TABLE 1.3 Astrobiological Methods and Interpretations

Does astrobiology require nonempirical methods?

Purpose of Astrobiology	*Purpose of Science*	
	Scientific Realism	*Constructive Empiricism*
Life	Yes	No
Evidence of life or of life-supporting conditions	Sometimes	No

In Table 1.3, we see that the constructive empiricist can reject nonempirical methods, because even if evidence of life or life-supporting conditions turn out to be unobservables, the constructive empiricist will simply hold that the purpose of science is not to find those either. But what if we are scientific realists? Insofar as astrobiology involves the search for life per se, then, astrobiology requires nonempirical methods. This does not entail that astrobiology requires *only* nonempirical methods, but a genuine discovery of life in this case will necessarily involve apprehension of the unobservable.

At the same time, we understand that astrobiologists do not search only for life per se, and this is especially true in the search for extraterrestrial life. For example, the search for biosignatures often involves searching for phenomena, such as liquid water or an atmosphere, that are both nonliving and capable of supporting life as we know it. And the possibility of finding extraterrestrial relics or other such evidence of life also engenders the search for that which is nonliving. When it was thought that Martian Meteorite ALH 84001 could potentially contain fossil evidence of past microbial life on Mars (Marchant, 2020), the appropriate response was empirical investigation. The crucial question then was not "Is it alive?" but "Is it a fossil?" Fossils are evidence of life, but they are not alive. And surely we would and should use empirical methods to find such evidence. Of course, some evidence of extraterrestrial life could, itself, be alive. Finding microbes on Saturn's moon Enceladus would mean we have found evidence of life, namely, life itself. Solely empirical methods will be insufficient here. So, for all astrobiological work that requires searching for and studying nonliving evidence and conditions, empirical methods are generally sufficient to conduct that fieldwork. But for all astrobiological work that involves searching for or studying living evidence and conditions, nonempirical methods will be needed under the assumption that scientific realism is true. Finally, there remains the question of whether and how we may confirm that something is alive—that it is a case of life per se. For that, nonempirical methods will be needed as well, given that life per se is an unobservable.

So, what does the above mean for astrobiology as it is practiced today? To date, we have detected life-supporting conditions in off-Earth settings but, as far as we can tell, have not detected life per se nor evidence of life per se. Thus, we have yet to encounter a scenario wherein nonempirical methods are needed to assess an ostensible instance of extraterrestrial life. Since there is much fruitful work to be done in both the search for life-supporting conditions that are not themselves alive, and in the search for evidence of life the evidence of which is not itself alive, our view allows that much of current astrobiological research can and should continue with business as usual and should continue to employ useful, empirical methods. Yet, foundational to astrobiological research is the goal of a confirmed detection of extraterrestrial life per se. This means that the nonempirical methods needed should be applied in two key ways: (1) they should apply to debates about which foundational assumptions belong in astrobiological research, and (2) they should be applied in certain instances of actual scientific practice, that is, in the event that we are in a position to attempt to confirm whether some extraterrestrial evidence is life per se. Given (1), there is reason here and now to apply nonempirical philosophical analyses to key conversations in contemporary astrobiological work.

Regarding (2), one might object by claiming that life's status as an unobservable prevents us from ever being certain that life exists, given that we cannot empirically confirm life itself. This implication of our view might seem implausible insofar as we are confident that we have confirmation of at least some life—that is, Earthly life. To respond, this confidence cuts both ways. Our confidence that life on Earth exists does not rule out the possibility that life is an unobservable, but it does provide some evidence that, if life is an unobservable, we may have confidence in the existence of unobservables. Likewise, we have confidence in the existence of time, metals, religion, and other such Group 3 phenomena. If the possibility that these too are unobservables does not appreciably weaken that confidence, neither should we assume that we should lack confidence in the claim that life exists here.

At the same time, we also wish to note that even those philosophers and scientists who seem to proceed under the perhaps-common assumption that life is the kind of thing that can be observed and is empirically verifiable allow that we are unlikely to achieve robust empirical confidence in the confirmation of extraterrestrial life in most, if not all, astrobiological searches. In a recent paper on astrobiology's empirical methods, Vickers et al. (2023) argue that we are even more limited than we typically think in our ability to legitimately express confidence in a candidate detection of extraterrestrial life and also to properly quantify our degree of uncertainty when it arises. Uncertainty is a significant problem, whether or not we are employing nonempirical methods.

Conclusion

One main role of the observable–unobservable distinction in the philosophy of science has been to require empiricists to explain our knowledge of unobservables or deny that there is such knowledge. In turn, if we think that science is largely successful, but we deny that there is knowledge of observables, then we must either deny the existence of unobservables or decide, with the constructive empiricists, that it is not really the point of science to pursue knowledge of unobservables.

Perhaps in some sciences, those options are acceptable. A quantum theorist may suggest that we simply "shut up and calculate" (Mermin, 2004), manipulating the pixels of macroscopic, observable computer monitors and recording the outcomes of macroscopic, observable measuring devices. This theorist may even deny that there is anything going on in an instrument when no measurement is being made.[34] Similarly, a social scientist may try to measure the cortisol levels of people who say that they attend church services at least once a week, prescinding from the question of whether the subjects of the research are members of a "religion." A cognitive scientist may record performance on some standardized test without, thereby, asking whether they are actually measuring the "intelligence" of the subjects.

Yet it is difficult to imagine a biology or an astrobiology that denies the existence of life or denies that the purpose of the field is, at least partly, to pursue knowledge of life. So, arguably, the point of astrobiology really is to study an unobservable. And, as we have argued, we should utilize nonempirical methods, such as logical argumentation, in order to evaluate which foundational assumptions about the existence of life and about knowledge of life itself belong in the field of astrobiology, and we must be open to revisiting and revising these assumptions in the face of new evidence, both empirical and nonempirical. We also ought to be prepared to evaluate claims and evidence of actual extraterrestrial life per se, and adequate preparation for the discovery of an unobservable can include nonempirical methods.

Therefore, we believe that astrobiologists, to the degree that they believe they are doing something important and worthwhile, have good reason to reject constructive empiricism. And if empiricism per se requires us to deny that we have knowledge of unobservables, then astrobiologists have good reason to reject empiricism per se. But it is not obvious to us that we must abandon empiricism in order to affirm the existence of unobservables. For example, one standard argument simply affirms that unobservables are the best explanation for the results we do observe, and that the practice of inference to the best explanation is perfectly consistent with empiricism, broadly construed. The presence of a living organism might be the best explanation for some measurements of gas in some atmosphere. Of course, constructive empiricists may reject inference to the best explanation (Monton and Mohler, 2023, § 3.2), at least as applied here, and this is not a chapter about constructive empiricism per se, so we cannot resolve that sub-debate. Yet we can at least see that our empiricist astrobiologist has a promising option available to affirm the existence of unobservables.

The other option is to simply reject any empiricism that is strict enough to forbid enlisting nonempirical methods and strategies. Such a strict empiricism is subject to several well-known and, in our opinion, powerful counterarguments anyway (BonJour, 1998). It is also generally rejected by experts.[35] For those of us who remain fascinated by the field of astrobiology, it makes much more sense to start by affirming the value and importance of astrobiology and seeing what that implies about our epistemology, rather than starting by imposing a dubious, procrustean epistemological assumption that might require us to abandon the search for life in the universe. Whether or not life is observable, it is certainly worth searching for.

Notes

1 A Humean view of causation (beyond constant conjunction) or persisting persons (beyond bundles of impressions or ideas) would be, roughly, that we have no sense impressions of them, so we are not justified in affirming their existence (Hume, 1976 [1740], 1.3.14.31; Penelhum, 1955). As for laws of nature, a Humean view would be that there is no impression of a law of nature beyond the specific phenomena that seem to be constantly correlated (cf. Lewis, 1986, p. ix; Caroll, 1990). See also Fodor (1989, p. 7) for examples.

2 For example, 72% of philosophers are scientific realists, and 64% of general philosophers of science join them. See PhilPeople (n.d.d).

3 It's mostly the non-realists who really need to rely on the distinction, because unobservables provide epistemological motivation for some forms of non-realism. See Chakravartty (2023, § 1.1). Then again, we will have to rely on the distinction as well, because we argue that the unobservability of life tells us other important things about philosophy and science.

4 See Kant (1998 [1781], A26, A33, A239).

5 This is obviously a deep and divisive topic. But it is traditional for empiricists to be skeptical of modal knowledge (indeed, this is a feature of debates over constructive empiricism [Monton and Mohler, 2023, § 3.5]), and empiricists strongly tend to reject platonism about abstract objects, see PhilPeople (n.d.b). Also, moral realists are more likely to accept platonism and to reject empiricism, see PhilPeople (n.d.c). There is substantial debate about whether these objects exist and are knowable, but much less debate about whether if they are knowable, that knowledge is nonempirical.

6 See n. 2.

7 Important court cases have centered on whether some group is a religion (Carolan, 2003). Physicists, chemists, and astrophysicists might all define "metal" differently (Swinburne University of Technology, n.d.).

8 For example, we may need to posit electrons to explain the behavior of cathode rays (American Physical Society, 2000), or conclude that moral facts exist because it is obvious that they do (Huemer, 2005), or because their existence explains the existence of moral progress (Huemer, 2016).

9 See, for example, Hasker and Taliaferro (2023, § 2). There might be further debate on whether the afterlife version of you is actually "alive," but we see no clear reason to fully reject the possibility. If we think there is no way to know whether the afterlife version of you is "alive," then, of course, that supports our contention that life is unobservable.

10 Some of these examples are reminiscent of the interesting case of chicken sexing, in which people seem to be able to perceptually distinguish features of objects without being able to consciously understand or describe how they are making these distinctions (cf. Horsey, 2002). We do not know whether chicken sexing is a form of empirically observing the sexes of the chickens, but as we discussed earlier, the existence of a few organisms that can observe some phenomenon empirically (for example, a mutant who could perceive radio waves) would not establish that the phenomenon is observable in general.

11 See BonJour (1985, p. 41 ff.) for this example in a mostly different context.

12 Although one might regard life as a natural kind (cf. del Savio, 2011), we wish to remain agnostic on the issue here, and we note that Group 3 unobservables include both natural kinds and things that are not natural kinds.

13 A mereological nihilist might say that tables and chairs don't exist (cf. Turner, 2011), but we may at least say that if they exist at all, they are observable.

14 A rare type of radical empiricist might believe in abstracta but believe that they are empirically observed. But there is a strong negative correlation between platonism about universals and empiricism (PhilPeople, n.d.b).

15 Of course, many historical descriptions or approximations of intelligence have been intentionally or effectively racist, sexist, ableist, or colonialist (cf. Reddy, 2008). However, this doesn't mean that it's not possible, in principle, to find some neutral and valuable idea of intelligence, nor does it imply that intelligence doesn't exist.

16 More generally, there is substantial debate about whether there is a unified definition of "religion." See Schilbrack (2023).

17 And there is no clear property that marks off metalloids from non-metalloids. See The Editors of Encyclopedia Britannica (n.d.).

18 See, for example, Hume (1976 [1740], p. 490) on language and money.

19 Contessa (2006) also provides a very useful discussion of these issues.

20 Arguably, some abortion law depends on a detection of life or life-related activity, for example, the "quickening" (Acevedo, 1979). While an embryo's being alive per se is rarely a key issue in the abortion debate among experts, popular discussions sometimes mention it (Pew Research Center, 2022).

21 For example, biocentrists believe that all living things have intrinsic value. See Taylor (1986) and Varner (1998).

22 If I wish to explain my persistence by positing psychological continuity, or by positing the existence of an animal or organism, then I may be relying on something like life (or something highly correlated with life) in order to separate persistence of nonliving objects from that of living objects. See for example Johnston (1987).

23 Design arguments seem most relevant. As an example, the standard fine-tuning argument holds that the universe is fine-tuned by some kind of designer for life (Collins, 2009; Waller, 2020). One potential problem for the fine-tuning argument is that the large-scale universe might be regarded as hostile to life, but if life is unobservable, then such a case may not be as straightforward to make.

Similarly, if life is unobservable, then we may not be as certain as we have been about the limits of what kinds of universe might permit life.

24 While many philosophers question the reliability of intuition or a priori methods, it is much less controversial that such methods are prominent and common in philosophical practice. See Climenhaga (2018) and Pust (2023, § 2).

25 Possibly, radical empiricists or naturalists would regard only observables to be relevant to philosophy (Mill, 1985, p. 125; cf. BonJour, 1998, ch. 3), and, probably, some philosophers of science or physics would regard knowledge of unobservables to be potentially very relevant to their philosophical work. But we doubt that most philosophers' day-to-day work would be affected much if they decided, for example, that microscopic particles were mere theoretical posits.

26 For example, affecting our brains may affect our moral judgments (Jeurissen et al., 2014). See Singer (2005) for a more general discussion.

27 One might consider the example of virtue epistemology here (cf. Zagzebski, 1996, 79 ff.).

28 See Baker (2023, § 1) for a sampling of appeals to simplicity in science.

29 For example, special relativity was partly derived from the principle of relativity (Einstein, 1912, p. 1059). One might also argue that an assumption of simplicity is necessary to rule out the possibility that forces other than gravity are affecting the path of light through space.

30 See Albert (1992, ch. 7) for a classic discussion of Bohmian mechanics.

31 See, for example, DeWitt (1970) for a landmark scientific statement and Greene (2011, ch. 8) for a popular account.

32 See Monton and Mohler (2023, § 3.10) for some discussion of whether constructive empiricism correctly captures the nature of scientific practice.

33 This conclusion won't surprise everyone who thinks about the epistemological foundations of science. For example, astrobiologists may appeal to inference to the best explanation, or to theoretical virtues such as simplicity or fecundity, in order to justify preferring hypotheses that posit the existence of life somewhere. Indeed, relying on inference to the best explanation is commonly taken to allow empiricist scientists to justify belief in unobservables (cf. Monton and Mohler, 2023, § 3.2).

34 "Shut up and calculate" is associated, in particular, with the Copenhagen interpretation of quantum mechanics (Mermin, 2004). Copenhagen theorists sometimes suggest that there is no point in asking what is actually going on with a system when it's not being measured (cf. Jammer, 1982, pp. 73–74; Omnès, 1994, p. 88).

35 A very strong majority of philosophers affirms the existence of a priori knowledge, thereby foreclosing the strictest forms of empiricism (PhilPeople, n.d.a).

References

Acevedo, Z. (1979). Abortion in early American. *Women & Health*, 4(2), pp. 159–167.

Albert, D. Z. (1992). *Quantum mechanics and experience*. Cambridge: Harvard University Press.

American Physical Society (2000) October 1897: The discovery of the electron. Available at: www. aps.org/publications/apsnews/200010/history.cfm (Accessed: 31 March 2023).

Baker, A. (2023). 'Simplicity', in E. N. Zalta (ed.), *The Stanford Encyclopedia of philosophy*. Winter edition. [Online]. Available at: https://plato.stanford.edu/archives/win2023/entries/simplicity/ (Accessed: 31 March 2023).

Bond, M. (2009, October 28). Clever fools: Why a high IQ doesn't mean you're smart. Available at: www.newscientist.com/article/mg20427321-000-clever-fools-why-a-high-iq-doesnt-mean-youre-smart/ (Accessed: 31 March 2023).

BonJour, L. (1985). *The structure of empirical knowledge*. Cambridge, MA: Harvard University Press.

BonJour, L. (1998). *In defense of pure reason*. Cambridge: Cambridge University Press.

Carolan, M. (2018, March 5). Scientology is 'a bona fide religion'. Available at: www.irishtimes.com/news/scientology-is-a-bona-fide-religion-1.350986 (Accessed: 31 March 2023).

Carroll, J. (1990). 'The Humean tradition', *The Philosophical Review* 99(2), pp. 185–219.

Chakravartty, A. (2023). 'Scientific realism', in E. N. Zalta (ed.), *The Stanford Encyclopedia of philosophy*. Winter edition. [Online]. Available at: https://plato.stanford.edu/archives/win2023/entries/scientific-realism/ (Accessed: 31 March 2023).

Churchland, P. M. (1985). 'The ontological status of observables: In praise of superempirical virtues', in P. M. Churchland and C. Hooker (eds.), *Images of science: Essays on realism and empiricism with a reply from Bas C. van Fraassen*. Chicago, IL: University of Chicago Press, pp. 35–47.

Cleland, C. E. (2019). *The quest for a universal theory of life: Searching for life as we don't know it*. Cambridge: Cambridge University Press.

Cleland, C. E. and Wilson, E. M. (2013). 'Lessons from Earth: Toward an ethics of astrobiology', in C. Impey, A. H. Spitz, and W. Stoeger (eds.), *Encountering life in the universe: Ethical foundations and social implications of astrobiology*. Tucson: University of Arizona Press, pp. 17–55.

Climenhaga, N. (2018). 'Intuitions are used as evidence in philosophy', *Mind* 127(505), pp. 69–104.

Cofield, Calla. (2018). "Search for extraterrestrial intelligence" needs a new name, SETI pioneer says. Available at: www.space.com/39474-search-for-extraterrestrial-intelligence-needs-new-name.html (Accessed: 31 March 2023).

Collins, R. (2009). 'The teleological argument: An exploration of fine-tuning in the universe', in W. L. Craig and J. P. Moreland (eds.), *The Blackwell companion to natural theology*. Chichester, West Sussex: Blackwell, pp. 202–281.

Contessa, G. (2006). 'Constructive empiricism, observability and three kinds of ontological commitment', *Studies in History and Philosophy of Science Part A* 37(3), pp. 454–468.

Del Savio, L. (2011). 'Life as a natural kind', *Logic and Philosophy of Science* 9(1), pp. 413–419.

De Witt, B. S. M. (1970). 'Quantum mechanics and reality', *Physics Today* 23(9), pp. 30–35.

Dicken, P. and Lipton, P. (2006). 'What can Bas believe? Musgrave and van Fraassen on observability', *Analysis* 66(3), pp. 226–233.

Einstein, A. (1912). 'Relativität und Gravitation. Erwiderung auf eine Bemerkung von M. Abraham', *Annalen der Physik* 343(10), pp. 1059–1064.

Fodor, J. A. (1989). *Psychosemantics: The problem of meaning in the philosophy of mind*. Cambridge: MIT Press.

Goldhill, O. (2016). Why are so many smart people such idiots about philosophy? Available at: https://qz.com/627989/why-are-so-many-smart-people-such-idiots-about-philosophy (Accessed: 31 March 2023).

Greene, B. (2011). *The hidden reality: Parallel universes and the deep laws of the cosmos*. New York: Alfred A. Knopf.

Hasker, W. & Taliaferro, C. (2023). 'Afterlife', in E. N. Zalta (ed.), *The Stanford Encyclopedia of philosophy*. Winter edition. [Online]. Available from: https://plato.stanford.edu/archives/win2023/entries/afterlife/ (Accessed: 31 March 2023).

Horsey, D. (2002). 'The art of chicken sexing', *UCL Working Papers in Linguistics* 14, pp. 107–117.

Huemer, M. (2005). *Ethical intuitionism*. Houndmills, Basingstoke, Hampshire: Palgrave Macmillan.

Huemer, M. (2016). 'A liberal realist answer to debunking skeptics: The empirical case for realism', *Philosophical Studies* 173(7), pp. 1983–2010.

Hume, D. 1976 [1740]. *A treatise of human nature*, 3rd ed. Oxford: Clarendon Press.

Jammer, M. (1982). 'Einstein and quantum physics', in G. Holton and Y. Elkana (eds.), *Albert Einstein: Historical and cultural perspectives: The centennial symposium in Jerusalem*. Princeton: Princeton University Press, pp. 59–76.

Jeurissen, D. et al. (2014). 'TMS affects moral judgment, showing the role of DLPFC and TPJ in cognitive and emotional processing', *Frontiers of Neuroscience* 8, art. 18.

Johnston, M. (1987). 'The problem of persistence', *Proceedings of the Aristotelian Society (Supplementary Volume)* 61, pp. 107–135.

Joyce, G. F. (1994). 'Foreword', in D. W. Deamer, G. R. Fleischaker (eds.), *Origins of life: The central concepts*. Boston, MA: Jones and Bartlett, pp. xi–xii.

Kant, I. (1998 [1781]). 'Critique of pure reason', in P. Guyer & A. Wood (eds.), *Critique of pure reason* (pp. 81–704). Cambridge: Cambridge University Press.

Lewis, D. (1986). *Philosophical papers*, volume II. Oxford: Oxford University Press.

Marchant, J. (2020). *Life on Mars: The story of meteorite ALH84001*. Available at: www.sciencefocus.com/space/life-on-mars-the-story-of-meteorite-alh84001 (Accessed: 31 March 2023).

Mariscal, C. (2023). 'Life', in E. N. Zalta (ed.), *The Stanford Encyclopedia of philosophy*. Winter edition. [Online]. Available at: https://plato.stanford.edu/archives/win2023/entries/life/ (Accessed: 31 March 2023).

Mermin, N. D. (2004). 'Could Feynman have said this?', *Physics Today* 57(5), pp. 10–11.

Mill J. S. (1985 [1840]). 'Coleridge', in J. M. Robson (ed.), *The collected works of John Stuart Mill, volume X: Essays on ethics, religion and society*. Toronto: University of Toronto Press, pp. 117–164.

Mix, L. J. (2015). 'Defending definitions of life', *Astrobiology* 15(1), pp. 15–19.

Monton, B. and Mohler, C. (2023). 'Constructive empiricism', in E. N. Zalta (ed.), *The Stanford Encyclopedia of philosophy*. Winter edition. [Online]. Available at: https://plato.stanford.edu/archives/win2023/entries/constructive-empiricism/

Muller, F. A. (2004). 'Can a constructive empiricist adopt the concept of observability?', *Philosophy of Science* 71(1), pp. 80–97.

Musgrave, A. (1985). 'Constructive empiricism and realism', in P. M. Churchland and C. Hooker (eds.), *Images of science: Essays on realism and empiricism with a reply from Bas C. van Fraassen*. Chicago: University of Chicago Press, pp. 196–208.

Normandin, S. and Wolfe, C. T. (eds.). (2013). *Vitalism and the scientific image in post-enlightenment life science, 1800–2010*. Dordrecht: Springer.

Omnès, R. (1994). *The interpretation of quantum mechanics*. Princeton, NJ: Princeton University Press.

Penelhum, T. (1955). 'Hume on personal identity', *The Philosophical Review* 64(4), pp. 571–589.

Pew Research Center. (2022). *Social and moral considerations on abortion*. Available at: www.pewresearch.org/religion/2022/05/06/social-and-moral-considerations-on-abortion/ (Accessed: 31 March 2023).

PhilPeople. (n.d.a). *A priori knowledge: Yes or no?* Available at: https://survey2020.philpeople.org/survey/results/4814 (Accessed: 31 March 2023).

PhilPeople. (n.d.b). *Abstract objects: Nominalism or platonism?* Available at: https://survey2020.philpeople.org/survey/results/4818 (Accessed: 31 March 2023).

PhilPeople. (n.d.c). *Meta-ethics: Moral anti-realism or moral realism?* Available at: https://survey2020.philpeople.org/survey/results/4866 (Accessed: 31 March 2023).

PhilPeople. (n.d.d). *Science: Scientific realism or scientific anti-realism?* Available at: https://survey2020.philpeople.org/survey/results/4910 (Accessed: 31 March 2023).

Pust, J. (2023). 'Intuition', in E. N. Zalta (ed.), *The Stanford Encyclopedia of philosophy*. Winter edition. [Online]. Available at: https://plato.stanford.edu/archives/win2023/entries/intuition/

Reddy, A. (2008). 'The eugenic origins of IQ testing: Implications for post-Atkins litigation', *DePaul Law Review* 57(3), pp. 667–678.

Schilbrack, K. (2023). 'The concept of religion', in E. N. Zalta (ed.), *The Stanford Encyclopedia of philosophy*. Winter edition. [Online]. Available from: https://plato.stanford.edu/archives/win2023/entries/concept-religion/ (Accessed: 31 March 2023).

Shaw, J. (2016). *Bill Nye and the value of philosophy*. Available at: www.rotman.uwo.ca/bill-nye-and-the-value-of-philosophy/ (Accessed: 31 March 2023).

Singer, P. (2005). 'Ethics and intuitions', *The Journal of Ethics* 9(3/4), pp. 331–352.

Swinburne University of Technology. (n.d.). *Metals. Cosmos: The SAO encyclopedia of astronomy*. Available at: https://astronomy.swin.edu.au/cosmos/m/Metals (Accessed: 31 March 2023).

Taylor, P. W. (1986). *Respect for nature: A new theory of environmental ethics*. Princeton: Princeton University Press.

The Editors of Encyclopedia Britannica. (n.d.). 'Metalloid', *Encyclopedia Britannica*. [Online]. Available at: www.britannica.com/science/metalloid (Accessed: 31 March 2023).

Turner, J. (2011). 'Ontological nihilism', *Oxford Studies in Metaphysics* 6, pp. 3–54.

van Fraassen, B. (1980). *The scientific image*. Oxford: Oxford University Press.

Varner, G., 1998. *In nature's interests? Interests, animal rights, and environmental ethics*. Oxford: Oxford University Press.

Vickers, P. et al. (2023). 'Confidence of life detection: The problem of unconceived alternatives', *Astrobiology* 23(7), pp. 1202–1212.

Waller, J. (2020). *Cosmological fine-tuning arguments: What (if anything) should we infer from the fine-tuning of our universe for life?* New York and London: Routledge.

Whitten, A. (2020). *Do IQ tests actually measure intelligence?* Available at: www.discovermagazine.com/mind/do-iq-tests-actually-measure-intelligence (Accessed: 31 March 2023).

Zagzebski, L. T. (1996). *Virtues of the mind: An inquiry into the nature of virtue and the ethical foundations of knowledge.* Cambridge: Cambridge University Press.

2

HOW TO THINK ABOUT UFOS

Robert CB Miller

> What does all this stuff about flying saucers amount to? What can it mean? What is the truth?
>
> (Winston Churchill)

Part I: Introduction – Toxic No Longer

Beginning in 2017, there has been a revolution in the treatment of UFO reports. On December 2017, articles appeared in the *New York Times (NYT)* and the *Washington Post (WP)* which explained that the American *Department of Defense* had spent $22m on a UFO research project called the *Advanced Aerospace Threat Identification Program (AATIP)* between 2007 and 2012. Since then, the project had continued without funding. The announcement was important for two reasons. First it came with the acquiescence (or perhaps even the support) of the US *Department of Defense (DoD)* which suggested that it (or at least part of it) believed that the phenomenon was real and genuinely puzzling. Second, the articles appeared in the *NYT* and the *WP* and both took the report at face value and did not subject them to ridicule or debunking. This is interesting as the *NYT* is famous for its fact checking and the *WP* cannot be far behind. It follows that one can take the reports seriously and the reader's natural scepticism should be restrained – if it is good enough for the *NYT* fact checkers, perhaps it should be good enough for you or me.

The thinking of the *DoD* appeared to be that, although UFOs did not seem to be a threat to national security, they had the capacity to be a threat. The reports included those from experienced US Navy pilots from the carrier USS Nimitz. Objects with the extraordinary flight characteristics of UFOs could operate as weapons against which the United States would have little or no defence.

DOI: 10.4324/9781003440130-4

UFOs used to be about as toxic in the 'academy' *as Intelligent Design* or *Young Earth Creationism*. But acceptance of UFOs as a legitimate field of scientific research was marked by an article in *Scientific American* in July 2020 by two scientists which asserted:

> Judging the nature of these objects (and these seem to be 'objects', as confirmed by the Navy) needs a coherent explanation that should accommodate and connect *all* the facts of the events. And this is where interdisciplinary scientific investigation is needed.

The article continued:

> As [Carl] Sagan concluded.... ..... 'scientists are particularly bound to have open minds; this is the lifeblood of science.' We do not know what UAP [i.e. UFOs] are, and this is precisely the reason that we as scientists should study them.
>
> *(Kopparapu and Misra, 2020)*

Further in June 2021, as mandated by an Act of Congress, the *Office of the Director of National Intelligence* published a report which confirmed the existence of such unexplained aerial phenomena and suggested that they might have multiple explanations (ODNI, 2021). Additionally, in 2023, NASA published a report that admitted the existence of phenomena which it could not explain (Spergel and Evans, 2023).

The result of these developments has been the strengthening of the evidence for the reality of the phenomenon. This has had several important consequences. The first is that UFOs appear potentially to be a genuine threat to air safety. The second is that the UFOs demonstrate powers unmatched by any planes of the US Air Force. In an interview on 20 July 2023 on ABC News, Dr Sean Kirkpatrick, the head of a US government task force set up to investigate the UFO phenomenon, the *All Domain Anomaly Resolution Office (AARO)*, stated that there was a danger of 'technological surprise' and that adversaries like Russia and China might achieve technological supremacy if the craft were theirs. Further there was also the alarming risk of 'extra-terrestrial technological supremacy' (ABC News, 2023). A further alarming consequence is what has been called 'ontological shock', resulting from the discovery of the existence of 'non-human intelligence' (NHI). This emerged during evidence (admittedly not eyewitness) that David Grusch gave to the *US House of Representatives Oversight Subcommittee* in July 2023. He claimed to have heard that the US government was in possession of crashed UFOs and their pilots. At the time of writing (September 2024), there is bipartisan pressure in both the House and Senate on the US government for the release of more information, and investigations by *AARO* and *NASA* continue.

Part II: UFOs – Evidence and History

Evidence and the Sagan ECREE Test

Perhaps the most important but least analyzed theme in the study of UFOs is what probative value should be given to UFO reports. There needs to be agreed reasonable grounds for evaluating the probative value of UFO reports. In other words, we need to have good grounds for deciding where credulity ends and credibility begins. Carl Sagan, who was a noted sceptic about UFOs (and religious belief), proposed the Extraordinary

Claims Require Extraordinary Evidence (ECREE) principle (Sagan, 1979/2011). This proposal was intended as criteria for ruling out rational belief in such things as ghosts, UFOs, and other supernatural beings and events. The principle can be compared to the 'moral certainty' or 'beyond a reasonable doubt' test used in criminal trials in common law countries. Its basic principle seems to be both reasonable and obvious. It is interesting that a similar principle was recommended by Herbert Thurston, a Jesuit priest, investigating the physical phenomena of mysticism (Thurston, 1952/2013, p. 2). Thus, I require better evidence to support belief in rare or curious claims than I do to believe reports of everyday events. For example, I would immediately believe a report of a van passing in the road next to my house, but a report of an elephant passing would provoke initial disbelief, investigation, accumulation of evidence, its assessment, and then a judgement on its credibility. The New Testament scholar, Richard Bauckham, has pointed out that exceptional events are better remembered than ordinary ones (Bauckham, 2017, 2018, pp. 55–71). Although the principle seems evident enough, it is difficult to give it precision. For example, what is to count as an extraordinary claim and what is to count as extraordinary evidence?

David Deming has pointed out that Carl Sagan's ECREE principle has a long history and originated in David Hume's famous essay *On Miracles* (Hume, 1748). Deming explains that the test of 'extraordinary evidence' is met by the reproducibility of evidence supporting a hypothesis (Deming, 2016). In the case of extraordinary events, this can be achieved by numerous independent eyewitnesses and recordings on several different sensors, all on several different occasions. One difficulty with the ECREE test is that it only shows which claims are belief-worthy, but not which claims are true. Thus, some claims may be true although the evidence for them may not reach the Sagan ECREE standard.

The UFO phenomenon is not all of a piece. The central case is where the evidence involves sightings of flying objects by military and civilian pilots, sometimes confirmed by numerous witnesses, film, and multiple sensors. Here the evidence is now very good and certainly passes the Sagan ECREE test. But there is a penumbra of other reports where the strangeness is far greater, abductions, crashed craft, their 'pilots', and the recovery and exploitation of 'alien' technology. Here it seems less plausible to claim that Sagan's ECREE test has been passed or even approached. Still the evidence for some of the strange events seems to be improving. For example, in the summer of 2023 a 'whistle-blower', David Grusch, emerged with the story that branches of the US government and private aerospace companies had numerous retrieved craft and the remains of dead 'pilots'. Plainly, as they stand, these reports do not even come close to passing the Sagan ECREE test. Such claims would include the Roswell UFO crash and 'pilot' retrieval of 1947 and, more recently, the 1996 Varginha UFO crash and retrieval case in Brazil. So extraordinary is the claim that the evidence must be super strong before belief in its truth is warranted. But Grusch, who formerly worked for *National Geospatial-Intelligence Agency (NGA)* and the *National Reconnaissance Office (NRO)*, served as the NRO representative on the *Unidentified Aerial Phenomena Task Force* from 2019 to 2021. He gave up his career and applied for 'whistle-blower' protection from the *Office of the Intelligence Community Inspector General (ICIG)* which judged his claims 'important and urgent'. None of this brings his claims close to meeting the Sagan ECREE test. Still, it shows progress towards meeting that standard, and further progress is possible. If this were to include photography and multiple reports by qualified eyewitnesses, then the Sagan ECREE test might be met.

This presents a problem in assessing the recent change in the quality of the evidence for the UFO phenomenon. As explained earlier, some of the evidence has greatly improved meeting the Sagan ECREE standard, while the evidence for others has improved but not to the Sagan ECREE standard. We will eschew such stories that include abductions, cattle mutilation, and personal contact with 'aliens'. These may be phenomena of a completely different kind, and in some cases, at least, they appear to be the result of rare psychological conditions, such as sleep paralysis.

Applying the Sagan ECREE Test

But how should the Sagan ECREE test be applied? It will be stiff but reasonable and appropriate to the circumstances and the phenomena. There will be multiple observers, radar returns from multiple sources, material from infrared cameras, and other sensors. For example, film was released of American fighter planes from the carrier USS Nimitz pursuing a UFO in the Pacific off San Diego. The object was described as about 40 foot in diameter and coloured white and when approached, it flew off at extraordinary speed. It also showed extreme manoeuvrability. No known aircraft is able to perform in such a way, and no living organism capable of controlling such an object could have survived the enormous g-forces – the equivalent of hitting a brick wall at several thousand miles per hour. The object gave a radar return, and it was witnessed by several flyers. The investigative journalist, Ross Coulthart, gives an excellent description of the incident (Coulthart, 2021, p. 136ff).

The pilots reported that the object had no control surfaces, and infrared cameras showed that they generated no significant heat. The Nimitz evidence confirmed that no heat is generated by their extraordinary acceleration and deceleration. Further, the object changed shape, and other evidence suggests that despite their enormous speed, they are virtually silent and do not create sonic booms. This report, and the excitement it generated, needs to be put into its historical and cultural context.

One difficultly is that the study of the phenomenon suffers from a mixture of extreme credulity and jumping to unwarranted conclusions – deriving a mountain range of mythology from a mole hill of fact. The mythology in its most developed form is gnostic, in that it refers to hidden knowledge, which is available to some of the governments of the world, which, for nefarious reasons, are keeping the truth to themselves. One supposed reason is that 'big oil' is alarmed at the ability of UFOs to fly without conventional fuels. Absurdities are common – and these include belief in multiple government conspiracies and secrets: theories about abductions, cattle mutilations, meetings between President Eisenhower and aliens, the existence of the Majestic 12 committee to supervise contact with aliens, and secret US government space programmes.[1] There is even discussion of a 'break-away civilization' (Dolan, 2016). Governments are also alarmed that the revelation of contact with aliens would cause a highly disruptive political and spiritual crisis with untold evil consequences. Such 'gnostic ufology' still seeks 'disclosure', and a retired A&E surgeon Dr Stephen Greer has organized Disclosure Events at the *National Press Club* in Washington. These had the virtue of assembling some very credible witnesses to some extraordinary aerial phenomena. But they are marred by the demand that the US government should disclose what they know about UFOs as the public has a right to know. What many ufologists fail to consider is that the US government (or rather some parts of it) may be just as puzzled as everyone else about the nature of UFOs and perhaps considerably embarrassed by that fact. The other mistake is to believe that governments are necessarily monolithic in their views. More often than not, they contain a variety of opinions about controversial (and not so controversial) subjects.

It is worthwhile to briefly analyze what is wrong with this approach to UFOs. Apart from a disgraceful and comic credulity, there is a tendency to believe that almost any explanation is better than none.[2] They are never content to admit ignorance. According to Wittgenstein (as described by his pupil and friend Con Drury), one of the roles of the philosopher is to show scientists the limit to their knowledge. Scientists (and others) often claim to know more than they do (Drury, 1973, p. 97). And there is often a strong case for freely admitting ignorance. And what is true of scientists is surely also true *a fortiori* of ufologists.

Early History

Much of what follows depends on Richard Dolan's comprehensive history of the UFO phenomenon between 1941 and 1991 (Dolan, 2002, 2009). Modern UFO history began in the WW2 with the reports by Allied aircrew of 'foo fighters', orbs of light which seemed to follow their planes. They were explained as a German secret weapon. German pilots reported similar sightings which were explained to them as Allied secret weapons (Dolan, 2002, p. 7). The next major incident and the start of the UFO craze proper of the 1940s and 1950s was the sighting by Kenneth Arnold in June 1947 of distant objects over Mount Rainier in Washington State, which he described as like 'saucers' skimming the surface of water. Thereafter followed a series of stories and incidents which have continued to this day. These include numerous stories of sky anomalies which are often, no doubt, the result of hoaxes, misidentifications of planes, drones, birds or obscure meteorology, and other natural events. Nonetheless there has been a residual where reliable witnesses, policemen, air force and commercial pilots, and other trained observers have given credible reports of strange objects in the sky.

Few public intellectuals have opined on the UFO question except to ridicule or to express scepticism about the reality of the phenomena. Typical is the case of Professor Carl Sagan (1934–1996) who was a confirmed and enthusiastic sceptic and took the view that all accounts could be explained as misidentifications or hoaxes. On the other hand, currently Professor Avi Loeb, a former head of the Harvard University astronomy department, has shown great interest in the phenomenon but tends to eschew explanations which include exotic science (Loeb, 2021; Kirkpatrick and Loeb, 2023).

In the 1950s, UFOs received the attention of the Carl Jung (1875–1961), who in 1958 published *Flying Saucers: A Modern Myth of Things Seen in the Sky* (Jung, 1958). He explained:

> Their [ie UFOs] progression in space is not in a straight line and of constant velocity like a meteor, but erratic like the flight of an insect and of varying velocity, from zero to several thousand miles per hour. The observed speeds and angles of turns are such that no earthly being could survive them any more than he could the enormous heat generated by friction.
>
> *(1958, p. 148)*

This account, which Jung based on contemporary reports (i.e. more than 60 years ago), is almost identical to those of the Nimitz fighter pilots. Their reports add little to Jung's summary of the evidence.

A strange feature of the UFO phenomena is that it only emerged during WW2. Prior to that date, there is minimal evidence for their existence. Retrospectively, ufologists have interpreted aerial vehicles in history, mythology, and art as UFOs, but these were not aerial sightings as they were only seen from the ground, and it is unclear what is being represented. In the 1890s, there were reports of mysterious airships in the Midwest and elsewhere in North America, and there was a report of a UFO spotted in England in 1916 (Weinstein, 2001).

Blue Book, the Condon Report, and Beyond

The character of UFOs has not changed since they first appeared in the 1940s. This is important as it suggests then that UFOs are not a cultural artefact and that they represent some real phenomenon.[3] In the same way that it has remained unchanged so has its regular interpretation – the explanatory dyad of prosaic explanation or 'aliens' has remained constant. As we shall see, there are more possible explanations than these.

The response of the US Air Force and the Department of Defense has also remained largely unchanged, reflecting the scientific culture of the time. In the 1940s, the US Air Force established a group to investigate the sightings, and this led to General Twining's memorandum of September 1947, which includes the following:

a. The phenomenon is something real and not visionary or fictitious.
b. There are objects probably approximating the shape of a disc, of such appreciable size as to appear to be as large as man-made aircraft.
c. There is a possibility that some of the incidents may be caused by natural phenomena, such as meteors.
d. The reported operating characteristics such as extreme rates of climb, maneuverability (particularly in roll), and motion which must be considered <u>evasive</u> when sighted or contacted by friendly aircraft and radar, lend belief to the possibility that some of the objects are controlled either manually, automatically or remotely.
e. The apparent common description is as follows:
    (1) Metallic or light reflecting surface.
    (2) Absence of trail, except in a few instances where the object apparently was operating under high performance conditions.
    (3) Circular or elliptical in shape, flat on bottom and domed on top.
    (4) Several reports of well kept formation flights varying from three to nine objects.
    (5) Normally no associated sound, except in three instances a substantial rumbling roar was noted.
    (6) Level flight speeds normally above 300 knots are estimated.

(Twining, 1947)

General Twining was an important figure in the *US Air Force*. At the time of writing the report, he was the commanding officer of the *Air Materiel Command*, which, amongst other things, investigated the aircraft of foreign (and potentially enemy) origin. Again, like Jung, Twining's statement remains an accurate description of the UFO phenomenon seen today.

Following the famous 1947 sighting by Kenneth Arnold of 'flying saucers' over Mt Rainer, there was a spate of other UFO reports. As a result, the US Air Force established Project Blue Book in 1952 to investigate the reports, and Professor Alan J Hynek, an astronomer, was appointed as the scientific adviser. He became convinced of the reality of the phenomena. However, the US Air Force was sceptical, and following a report by Professor Condon of Colorado State University, which doubted the actuality of the phenomenon, the project was closed in 1969. And scepticism was general amongst science popularizers like Martin Gardner (Gardner, 1957), Carl Sagan, and Patrick Moore. The Condon report was widely criticized for not taking the evidence seriously, and Hynek remained a committed believer in the reality of the phenomenon, as explained in his book, *The UFO Experience: A Scientific Study* (Hynek, 1972). There matters rested, with US Air Force claiming that there was nothing that needed investigation until 2017. Despite this official indifference, there was a steady stream of credible reports collected by private agencies such as *Mutual Unidentified Flying Objects Network (MUFON)* in the United States.

From AATIP to AARO 2012–2023

In 2017, it emerged that the US government had launched a project, *the AATIP*, prompted by Senator Harry Reid with a funding of $20m. A leading light in the project was a career intelligence officer, Luis Elizondo, who resigned from government service to press the case for more UFO research. In 2017, he helped the pop singer Tom de Longe to set up a private UFO investigative group, the *To the Stars Academy*. More importantly, experienced US Navy pilots, David Fravor and Ryan Graves, gave evidence supported by multiple witnesses and sensors of UFOs with extraordinary flight characteristics. As we have seen, this led, in turn, to the formation of *AARO*, bipartisan Congressional pressure, and hearings before Congress, which included David Grusch's extraordinary testimony and confirmatory statements by the US Navy aviators, David Fravor, and Ryan Graves – all under oath.

Part III: UFO/UAP Characteristics

UFOs have a number of significant characteristics. One important development since the 2017 advent of the new ufology has been the description of the 'five observables' by the analyst Luis Elizondo who was employed by *AATIP*. These five are described below (1–5), along with four other important characteristics (6–9). They...

1 ...show anti-gravity lift without wings or other sources of lift.
2 ...show sudden and instantaneous acceleration. Such extraordinary rates of acceleration and deceleration would destroy craft or biological entities.
3 ...show hypersonic velocities without signature. In other words, they show extraordinary speed without a sonic boom, exhaust or friction with the atmosphere.
4 ...have low observability or cloaking. In other words, they appear able to appear and disappear at will.

5 ...demonstrate trans-medium travel. This means that thy can transit from below the sea (unidentified submerged objects [USOs]) to the air and from the air to space.
6 ...take numerous different forms. UFOs come in many different shapes and sizes. Shapes range from lights to enormous aircraft carrier-sized objects. Shapes include triangles, lights, circles, disks, ovals, cigars, and fireballs.
7 ...seem not to employ radar and they neither send nor respond to radio transmissions – no radio message has ever been received from a UFO.
8 ...show particular interest in nuclear power or nuclear weapon facilities (Hastings, 2017).
9 ...UFOs display intelligence. They display inquisitiveness, 'pace' aircraft, show evasiveness, and, on rare occasion, defensiveness. When observed, they will fly off at very high speed.

What are UFOs Doing?

Given that they evidently act with intelligence and purpose, it raises the question of what they are doing. Their actions, while plainly intentional, are 'brute' in the sense that they admit of no complex description (Anscombe, 1958; Geach, 1969, p. 31). They have been seen to 'pace' aircraft military and civilian, fly over missile silos, or flit over the peaks of an Antarctic mountain range. Jung compares their actions to those of insects, but this seems merely to describe their sudden changes of direction. The movements of insects can always be fitted into some story relating to their form of life. But UFOs seem to have no form of life – or even life as far as we can recognize it. There are no reports of them feeding, excreting, mating, reproducing, or giving birth. Nor is there evidence that they are artefacts, although this is often assumed. There is no evidence of their construction, destruction, fuelling, being serviced, or dumping waste products. They evidently have an equivalent of animal intelligence, but it is wholly unclear what they are doing. What they are doing is as puzzling a question as to what they are. Some may be 'manned', others (perhaps the vast majority) may not be. And some may be remotely controlled drones or even van Neumann self-replicating probes from a distant star system.

Part IV: Explaining the UFO Phenomenon

Before reviewing the possible explanations of the phenomenon, it is worth countering four popular fallacies.

The Four Fallacies of Ufology

The first is the *explanatory bifurcation fallacy*. It is the assumption that there are only two possible explanations for the UFO phenomenon: either it has a prosaic natural explanation, such as a misidentified experimental aircraft or atmospheric phenomena, or it is an extraterrestrial craft. But this, as will be discussed below, is to limit the number of possible explanations. An egregious example of this fallacy is to be found in NASA's 2023 report on unidentified aerial phenomena (UAP) (Spergel and Evans, 2023).

The second is the *necessary technology fallacy* which supposes that the extraordinary powers of UFOs are the result of technology. But while this may be true, it is not the only explanation. Thus, the power of birds to fly is not the result of technology, and it may be that UFOs are more like birds than they are like planes. That they have pilots needs to be demonstrated and not just assumed.

The third fallacy is that despite its diverse character, the phenomenon has only one explanation, the *single explanation fallacy*. Given this diversity described earlier, this seems highly implausible. For example, one possibility is that many UFOs are experimental aircraft of the US government. Others may have other explanations. It may be that explanations for some UFOs will be discovered, but others may remain mysterious.

The fourth fallacy is the *fallacy of dogmatic prosaicism*. This is the belief, assumed but rarely explicitly justified, that if no prosaic explanation can be found, then the absence of such an explanation is the result of a shortage data. But as the ufologist Stanton Friedman pointed out, often the most puzzling cases are those where the evidence is good. It may be the case that some puzzling reports may be resolved prosaically, but this is not necessarily the case. An illustration of this error is in a paper 'Physical Constraints on Unidentified Aerial Phenomena' by Harvard astronomer Avi Loeb and Sean Kirkpatrick (2023), the head of AARO. They derive '...physical constraints on interpretations of "highly manoeuvrable" UAP based on standard physics and known forms of matter and radiation'. They then suggest reports of phenomena outside these limits likely suggest inaccurate measurements (Kirkpatrick and Loeb, 2023, p. 1). While this seems a sensible approach, it has the implication that any observation outside these limits *must* be the result of inaccurate reports.

The Eight Possible Explanations of UFOs

But how can UFOs be explained? Given the general acceptance of a scientific view of the world, some explanations may be more or less congenial depending on the position taken on such issues. The eight possible explanations analyzed below appear to be a comprehensive list.

Prosaic Natural Explanations

Prosaic natural explanations remain the preferred account of the people with a scientific world view. UFOs are explained as illusions, hallucinations, misinterpreted astronomical or meteorological phenomena, drones, balloons, birds, or secret military aircraft. In particular, some reports of delta-shaped UFOs may well represent secret military aircraft. While most UFO reports can be explained prosaically, there is a residual number where no prosaic explanation is possible. The difficulty is that the evidence for these residual cases seems to pass quite a demanding version of the Sagan ECREE test. Still this view is still pervasive amongst scientists which explains why no articles about UFOs are to be found in *Nature* or *Science* – or any other mainstream scientific journal. (But as we saw earlier, the exception is the July 2020 article in *Scientific American*.) Some UFOs are evidently the secret aeroplanes of governments (American and possibly those of other countries). But given their extraordinary flight characteristics, described earlier, it seems unlikely that this can be true of all of them. Another possibility is that some, at least, might represent the product of a US government psychological operation or 'psyop' for some purpose, perhaps to discredit reports of genuine secret aircraft.

Extraterrestrial Hypothesis

The extraterrestrial hypothesis (ETH) is that UFOs are craft from another solar system, which are likely 'manned' by alien crew. But this account, which is the second horn of the *explanatory bifurcation fallacy*, has some serious difficulties. In the first place, there is the problem of distance, and the time and energy needed to travel from a faraway star system. UFOs never appear to refuel or to discharge waste. Further it is completely unclear why UFOs would come all this way without an obvious reason. The solution proposed for these difficulties is that the UFOs or their pilots use 'advanced' technology that allows them to travel about the universe with ease and speed. It also explains their extraordinary flight characteristics. Still, it leaves unresolved the puzzle of what UFOs are doing. They carry out simple 'brute' actions: following, chasing, being inquisitive, and pacing. But they can never be put into a more complex framework of activity. Much effort has been given by organizations such as *The Scientific Coalition for UAP Studies (SCU)* to developing exotic scientific theories to explain their behaviour. But such theories are not accepted by mainstream science, and even if shown to be plausible, *they could not account for the existence of UFOs or what they were doing.* The theories assume that UFOs are tractable to science, but this common assumption may be mistaken. Still even if these theories could be shown to be true, it would not explain their nature, motives, or origin.

Time Travel Hypothesis

It has been suggested that UFOs are human travellers from the distant future, and that the strange human-like forms of these time travellers are the result of the future evolution of the human species. This theory advanced by Dr Michael Masters has a number of obvious difficulties (Masters, 2019). In the first place, backwards time travel (in particular) is replete with paradox. For example, a time traveller from the future would be able to change what (for it) was the past and in principle prevent the birth of the time traveller – the grandfather paradox. Another serious difficulty is that Masters takes too seriously those who seek to predict the future course of human evolution.

Alternative Life Form

One explanation is reminiscent of a story by Sir Arthur Conan Doyle which appeared in the *Strand Magazine* in 1913. *The Horror of the Heights* (Doyle, 1913) told how early aviators flying very high had mysteriously disappeared. The explanation was that they had been attacked by flying creatures which destroyed them and their machines. The creatures living above the clouds at great height formed an ecosystem of their own. The story is interesting for two reasons. First, it seems to anticipate UFOs three decades before the UFO reports by German and Allied pilots in WW2.[4] Second, it offers an alternative to the 'illusion or aliens dichotomy', which was described earlier as the first of the four fallacies of ufology. Whether UFOs could be a kind of flying animal must be very doubtful given the extraordinary flight characteristics of UFOs, which, as we have seen, appear not to be affected by gravity or inertia. Nor is there any evidence of the natural ecology which any kind of animal would have. But one virtue of the possibility raised by Conan Doyle is that it captures the purposeful behaviour of UFOs. Conan Doyle's story is important as it suggests solutions to the UFO puzzle which are excluded by the usual dichotomy of illusion or alien visitors. One difficulty is that they display considerable intelligence *but without ecology*. As we have seen, they never refuel or emit exhaust. Nor are they ever seen to hunt, feed, or reproduce.

Recently, though, the philosopher Bernardo Kastrup has argued that UFOs may be the 'remnants of industrial technology NHIs evolved on earth up to 350 million years ago' (Kastrup, 2024). The obvious difficulty with this 'ultraterrestrial' theory that there is no geological or archaeological evidence of such a civilization is explained by the fact that weather erosion and the regular recycling of the earth's crust by plate tectonics would have destroyed any evidence long ago. He claims reasonably enough that if any 'biologics' retrieved had the same DNA as life on earth, it would strengthen the case for this explanation (Kastrup, 2024). But without any positive evidence, the theory seems unduly speculative (even in a field where wild fancy is all too common).

Multidimensional Hypothesis

The doyen of ufologists, Jacques Vallee, has claimed in a number of books that UFOs are a kind of quasi-religious phenomenon. In the same way that, in the past, the presence of angels, demons, and fairies was almost universally accepted, so these mythical beings are seen today as UFOs (Vallee et al., 1965, 1969, 1975, 1988, 1991). This view is similar to that of the psychologist Carl Jung, whose book with its telling title, *Flying Saucers: A Modern Myth of Things Seen in the Sky*, is an early example of this explanation (Jung, 1958). More recently, the theory has also been advanced by a professor of comparative religion, Diana Pasulka (Pasulka, 2019). UFOs are, thus, to be explained as entities from other dimensions which intrude on humanity and are the source of religious and quasi-religious ideas. For example, Vallee interprets the Fatima phenomena as similar to that of UFOs (Vallee, 1975/2014, pp. 141–154). In the same way that a three-dimensional object passing through a two-dimensional plane world would first appear and then disappear, so UFOs are supposed to move through our world from one (or more) dimensions. There are numerous difficulties with this view. There is no strong independent evidence that such extra dimensions exist. Extra dimensions (to a total of ten) are theorized by string theory, but beyond the standard three, these are wrapped up and are inaccessible to us. Another difficulty is that Vallee gives credence to a large number of events and 'cases' which do not come even close to passing the Sagan ECREE test. In many cases, they appear to be the result of psychological phenomena, such as sleep paralysis.

Spirits of the Abrahamic Religions Theory

If one has a world view, which accepts the existence of an unseen spirit world as asserted by the Abrahamic religions, another explanation is available – that UFOs are spirits, or similar to spirits. Those who do not share this world view can disregard the remainder of this section.

Spirits and angels are found in both the Hebrew Bible and the New Testament and in the Qur'an (Jones, 2011; al-Ashqar, 2020). It is the Angel Gabriel who imparts the Qur'an to Mohammed by whispering in his ear. It is important to distinguish between spirits and angels. The latter are spirits with the role of messenger. All angels are spirits but not all spirits are angels. Some spirits are fallen angels or demons. Hugh Ross, an evangelical Christian apologist, has even argued that UFOs are demons because aliens are likely to be extraordinarily rare (if they exist at all), and it would be very difficult for aliens to travel to Earth (Clark, Ross, and Samples, 2004). In Islamic theology, there are entities known as the jinn, which seem to occupy an intermediate position between angels and demons. Jacques Vallee has pointed out that traditional European folklore included stories of fairies, and these are similar in many respects to reports of UFOs. In his book, *Passport to Magonia* (Vallee, 1969/2014), Vallee cites the 17th-century clergyman Robert Kirk's description of fairyland in his *The Secret Commonwealth of Elves, Fairies and Fauns* (Kirk, 1691[?]/1815/2023). In his introduction, Andrew Lang explains how Kirk believed that 'the land of fairy was a fact of nature' (Kirk, p. 6).[5]

Aquinas's analysis of the nature of angels shows the attractions of this explanation. He argued that spirits (including we may suppose fairies and jinns) are pure forms without matter. This follows his basic metaphysical distinction (derived from Plato and Aristotle) between matter and form. Matter is the basic stuff out of which things are composed, and their form is their configuration or structure. Human beings are composed in the same way, except that their form is self-subsistent because it is rational. Angels also are rational and are created directly by God. They are immortal because of their character as 'self-subsistent forms' or pure configuration, and there is a sense in which they are not alive like an animal or a human being. Nor strictly do they exist in spatial dimensions although they can appear to us. They do not eat or digest food and do not reproduce. Angels, Aquinas claims, 'assume bodies of air, condensing it by the Divine power insofar as is needful for forming the assumed body' (Aquinas, Art 2, Reply to Objection 3). They have the power to cloak themselves with matter as and when they choose.

Aquinas was puzzled by the cognitive powers of angels which do not have sensory organs, such as eyes and ears, and he argues their knowledge derives from pure intellection.[6] This question reminds one of the occasional inability of UFOs to detect planes when they themselves have been detected.

How does this description of angels illuminate the character of UFOs? This can be answered simply – *they have similar powers*. As spirits are not material objects, they are not subject to the limitations of biology or physics – they can survive any acceleration or deceleration, and they can move very sharply in any direction. Because they can take on any material form, they can change shape at will, and because they are pure intellect, they can demonstrate purpose and intelligence. In addition, their character as pure form explains why they have such limited effects on their surroundings; their movements do not cause wake or turbulence in the air behind them as they move at extraordinary speed.[7] In other words, they are not subject to the limitations of the laws of physics. It is intriguing too that angels like UFOs are creatures of the air – that is why the former are usually portrayed with wings.

It follows that their character as pure intellect explains how they can both travel at extraordinary speeds and just disappear from vision and radar return. Their lack of mass accounts for their quietness and the absence of sonic booms when they accelerate. The peculiarity of their cognitive powers (by human standards) may explain how they are often observed before they detect that they have been seen. Planes have to get quite close before they take evasive action. It would also account for their lack of technology. For example, they never use or display weapons or any other tools. Witnesses, most often air crew, are left puzzled but only occasionally damaged or fundamentally troubled.

UFOs often (but far from universally) appear indifferent to humans. It does not appear that UFOs are fallen angels or evil spirits; they seem to display neither malice nor beneficence. This does not fit well with the conventional division of spirits into benevolent angels and malevolent demons. There are no stories of UFOs preventing air crashes in the same way that dolphins have been known to save people from drowning. As we have suggested, there may be a third class: betwixt and between. It follows that the character of fairies or jinns in many respects fits well with the perceived character and activities of UFOs.

Chomsky/McGinn Mystery

One approach is to adopt a quasi-scientific position and argue that there is a residual where satisfactory natural explanations are impossible. The phenomenon may be of a type for which we cannot find an explanation, and that it may remain mysterious forever. This view, whose supporters are known unkindly as 'mysterians', is that human nature is such that we can pose questions we are constitutionally unable to answer. In the same way that a rat cannot ponder the progression of the prime numbers, so some questions, typically the puzzle about the freedom of the will, may not be soluble by human beings. This view is supported by the linguist Noam Chomsky and some of his followers such as the philosopher Colin McGinn. This view represents an attempt by Chomsky, in philosophical mode, to combine a naturalistic approach to human nature without ruling out the reality of our experience of freedom. Mysteries of this sort seem to be the result of antinomies, such as the reconciliation of physical law and freedom. McGinn contrasts mysteries which are insoluble because of the weakness of our intellects with puzzles which are soluble in principle (McGinn, 1993). (There is no suggestion that either Chomsky or McGinn applied this doctrine to the UFO phenomenon.)

Chomsky/McGinn mysteries are similar to the sacred mysteries described in Christian theology. Belief in such mysteries is part of Catholic doctrine which holds that human reason is incapable of penetrating the full meaning and significance of, for example, the Trinity, the Virgin Birth of Jesus, and the Resurrection. Like Chomsky/McGinn mysteries, these theological mysteries derive from the weakness of the human intellect.

Lewontin Puzzle

Another possibility is that investigations cannot be successful, and the phenomenon may remain unexplained, possibly forever. The famous biologist R. C. Lewontin was asked to contribute an essay on the evolution of human cognition. His first response was to reject the invitation as nothing sensible could be said on the subject. But being pressed further, he contributed a famous article: *The Evolution of Cognition: Questions We Will Never Answer*, in which he argued that it would never be possible to give an account of the evolution of human cognitive powers, because the necessary evidence no longer existed as it had been destroyed long ago (Lewontin, 1998). Theories, 'just-so-stories', might be developed, but deciding which, if any, were true would always remain impossible. In a similar way, the explanation of the UFO phenomenon may remain hidden because the necessary evidence may not (and may never) be available to us.

Part IV: Ontological Shock – an *After Grusch (AG)* **World**

Parts II and III of this paper have analyzed the UFO phenomenon on the assumption that some of the evidence currently passes the Sagan ECREE test, but other reports and claims do not. In what follows it is assumed that the extreme claims of the 'whistle-blower' David Grusch are shown to pass the Sagan ECREE test. These claims included the following: The US government has in its possession some crashed (or abandoned) craft which are of non-human origin, together with some of the occupants. The craft are being investigated with a view to 'back-engineering' their extraordinary powers. It goes without saying that these hyper-extraordinary claims have yet to be backed by evidence close to meeting the Sagan ECREE standard. But it must be admitted that David Grusch's assertions are far better than the febrile speculation about crash retrievals that existed before and that was so easily (and rightly) dismissed.

What would count as meeting the ECREE standard for such hyper-extraordinary claims? The evidence would have to include multiple independent eyewitnesses, photographs, and reports by independent scientists. Now suppose that the evidence meets the Sagan ECREE standard, what would we think? In the first place, we should attempt to come to terms with the 'ontological shock' (a term Grusch used) that the evidence would produce. It would amount to a grave affront to the modern scientific view of the world. The evidence would resolve the often-posed question: 'Are we alone in the universe?' By this it is meant that we are not the only intelligent creatures in the universe. This would challenge our modern scientific view. But 500 years ago, the question would have been dismissed as nonsensical as belief in witches, fairies, angels, demons, and jinns was common, if not universal. It follows that the AG world would be a reversion to something similar to a pre-Enlightenment worldview. Belief in fairies and witches was dismissed as 'old wives' tales', but maybe the old wives were onto something.

In an AG world, one important topic of research would be whether the UFOs and their crew were supernatural entities, objects subject to the known laws of physics or subject to laws which have not yet been discovered. Or indeed whether they originated from another star system. Strong evidence would be required to make this credible. One consequence of the recovery of craft and their 'pilots' would be that it becomes less likely that all (or even some) UFOs are spiritual entities, or indeed entities from another dimension or dimensions. The recovery of craft and pilots would strengthen the case for the ETH; UFOs are just the craft of aliens from a different star system visiting the earth for unknown reasons. Their materiality might weaken the Multidimensional Hypothesis, and the spirits of the Abrahamic religions theory would become less credible. If any UFO occupants had the same DNA as other life on earth, the alternative life form theory would be strengthened. If their biology was based on a wholly different biochemistry, this would further strengthen the ETH. All varieties of prosaic explanations would be rendered implausible. The Lewontin Puzzle and Chomsky/McGinn Mystery theories would remain intact but weakened.

Part V: Conclusion

David Grusch's use of the phrase 'ontological shock' in the discussion about the UFO phenomenon is important. By this phrase he seems to have meant that the UFO phenomenon represents an affront to the modern scientific view of the world.

Maybe we do live in a universe which includes human beings, angels (fallen and otherwise), and UFOs as supernatural entities – amongst several others.[8] In the Enlightenment, there was a natural desire to reject all superstition.

> The eighteenth century was not a period notable for scientific advance. It was, however, an age profoundly affected by the scientific outlook. The influence of Newton was all pervasive. ..... The results of this new attitude appeared in a variety of ways. The authority of science combined with the prestige of reason to challenge every form of superstition; traditional views were suspect till they had been scrutinised and tested.
>
> *(Cragg, 1964)*

And this approach involved using very strong rules of evidence. Remember how the *French Academy of Sciences* refused to countenance reports of meteorites until confronted with the very clearest evidence. Maybe we have been doing the same thing about UFOs. Like the French academicians we may be missing something very interesting. One possibility is that there is an extreme type of Kuhnian paradigm shift in the offing. What form this revision of the modern world view may take is unclear. It will surely depend on the evidence that emerges. If good evidence for crash retrievals appears, then the conceptual revision will be (substantially) different from what is to be expected if the evidence turns out to be weak or non-existent. Another possibility is that the phenomenon is intractable to any form of investigation, and that no relevant evidence ever emerges. We may just be left with a puzzle or a mystery. Winston Churchill's question, posed more than 70 years ago, is no closer to being answered now than it was in the 1950s. For the present, we may be left just puzzled or mystified.

Glossary

AARO – All Domain Anomaly Resolution Office
AATIP – Advanced Aerospace Threat Identification Program
UFO – Unidentified Flying Object; UAP – Unidentified Aerial / Anomalous Phenomenon
USO – Unidentified Submerged Object.

Notes

1 The Majestic 12 organization was set up supposedly by President Truman in 1947 to supervise UFO crash retrievals. The documents on which this belief is based are widely believed to be forgeries.
2 One exception was the grandfather of Ufology, Stanton Friedman (1934–2019), who retained a healthy scepticism about most UFO reports.
3 I owe this point to Mr. Richard Miller.
4 It is puzzling that there is only one report of a UFO being seen during WW1 despite considerable aerial activity over the western front (Weinstein, 2001).
5 Belief in fairies persisted for a long time. As a child in Belfast, C. S. Lewis had a nurse from the Mountains of Mourne in Co Down who claimed to have seen the fairies.
6 The story is told that when Aquinas had a heavenly vision, his first response was to ask as to the means of knowledge that angels had.
7 Aeroplanes reporting to Air Traffic Control describe themselves as, say 'Speedbird 637 Heavy', meaning that the plane is a large (i.e. 'heavy') BA airliner which will cause turbulence that could interfere with the flight of planes flying close behind it. Imagine the turbulence caused by a conventional object flying at multiple thousand miles an hour at low altitudes. UFOs never seem to have this or indeed any other effect. And they have never been known to cause sonic booms.
8 I owe these thoughts to Mr. Richard Miller.

References

ABC News, 2023. *Pentagon UFO Investigator: Extraterrestrial 'Technical Surprise' Is Top Concern | ABCNL*. [Online Video] Available at: www.youtube.com/watch?v=ifpLXP0poug (Accessed 10 May 2024).

al-Ashqar, U., 2020. *The World of the Jinns and Devils*. International Islamic Publications House.

Anscombe, G. E. M., 1958. 'On Brute Facts'. *Analysis*, Vol 18, No 3, pp. 69–72.

Aquinas, 'Question 51. 'The Angels in Comparison with Bodies' *Summa*, Online Edition. (Accessed January 2018).

Bauckham, R., 2017 (First published 2006). *Jesus and the Eyewitness*. Grand Rapids: Wm. B. Erdmans Publishing Co.

Bauckham, R., 2018. 'The Gospels as Testimony to Jesus Christ: A Contemporary View of Their Historical Value'. In F. A. Murphy (Ed.), *The Oxford Handbook of Christology*. Oxford: Oxford University Press, pp. 55–71.

Clark, M., Ross, H., and Samples, K. 2004. *Lights in the Sky and Little Green Men*. Illinois: NavPress Publishing Group.

Coulthart, R., 2021. *In Plain Sight*. Sidney: Harper Collins.

Cragg, G. R., 1964. *Reason and Authority in the Eighteenth Century*. Cambridge: Cambridge University Press.

Deming, D., 2016. 'Do Extraordinary Claims Require Extraordinary Evidence?'. *Philosophia*, Vol 44, pp. 1319–1331.

Dolan, R. M., 2002. *UFOs and the National Security State: Chronology of Cover-Up 1941–1973*. Charlottesville: Hampton Roads Publishing Company Inc.

Dolan, R. M., 2009. *UFOs and the National Security State: The Cover-Up Exposed 1973–1991*. Charlottesville: Hampton Roads Publishing Company Inc.

Dolan, R. M., 2016. *The Secret Space Program and the Breakaway Civilization*. Richard Dolan Press.

Doyle, A. C., 1913. 'The Horror of the Heights', *The Strand Magazine*, November.

Drury, M. O'C. 1973. *The Danger of Words*, London and New York: Routledge and Kegan Paul.

Elizondo, L. 2018. 'Exclusive 60 Minutes with Luis Elizondo – Former Director of the AATIP' (UFORadio International #11) Available at: https://youtu.be/zhvClGiBB2c (Accessed 12 February 2018).

Gardner, M., 1957. *Fads and Fallacies in the Name of Science*. New York: Dover Publications.

Geach, P. T., 1969. *God and the Soul*. London: Routledge and Kegan Paul.

Gilson, E., 1945. *Le Thomisme*. Paris: Librairie Philosophique J. Vrin.

Hastings, R., 2017. *UFOs and Nukes*. Self-published.

Hume, D., 1748. 'On Miracles'. *Philosophical Essays Concerning Human Understanding*. London: A. Millar.

Hynek, J. A., 1972. *The UFO Experience: A Scientific Inquiry*. Chicago: Henry Regnery Company.

Jones, D. A., 2011. *Angels: A Very Short Introduction*. Oxford: Oxford University Press.

Jung, C. G., 1958. *Flying Saucers: A Modern Myth of Things Seen in the Sky*. London and Henley: Routledge and Kegan Paul.

Kastrup, B., 2024. 'UAP and Non-Human Intelligence: What Is the Most Reasonable Scenario?'. *The Debrief*, 6 January. Available at: https://thedebrief.org/uaps-and-non-human-intelligence-what-is-the-most-reasonable-scenario/#sq_hlkk65yczy (Accessed 10 May 2024).

Kean, L. 2010. *UFOs*. New York: Three Rivers Press.

Kirk, R., 1691(?)/1812/2023. *The Secret Commonwealth of Elves, Fairies and Fauns*. Loschberg: Jazzybee Verlag Jurgen Beck.

Kirkpatrick, S. and Loeb, A., 2023. 'Physical Constraints on Unidentified Aerial Phenomena'. Available at LK1.pdf (harvard.edu) (Accessed 21 September 2023).

Kopparapu, R. and Haqq-Misra, J., 2020. '"Unidentified Aerial Phenomena," Better Known as UFOs Deserve Scientific Investigation', *Scientific American*, July 27.

Lewontin, R. C., 1998. 'The Evolution of Cognition: Questions We Will Never Answer'. In Scarborough, D. and Sternberg, S. (Eds.), *An Invitation to Cognitive Science, Volume 4: Methods, Models, and Conceptual Issues*. Cambridge: MIT Press, pp. 107–132.

Loeb, A., 2021. *Extraterrestrial*. London: John Murray.

Masters, M. P., 2019. *Identified Flying Objects*. Self-published.

McDonald, J. E., 1969. 'Science in Default'. *American Association for the Advancement of Science*, 134th Meeting. Available at: http//dewoody.net/ufo/Science_in_Default.html (Accessed March 2018).

McGinn, C., 1993. *Problems in Philosophy the Limits of Inquiry*. Oxford and Cambridge: Blackwell.

ODNI, 2021. *Preliminary Assessment: Unidentified Aerial Phenomena*. Washington: Office of the Director of National intelligence.

Ott, L., 1955/1963. *Fundamentals of Catholic Dogma*. Cork: The Mercier Press.

Pasulka, D. W., 2019. *American Cosmic*. Oxford: Oxford University Press.

Sagan, C., 1979/2011. *Broca's Brain: Reflections on the Romance of Science*. New York: Random House.

Spergel, D., Evans, D. et al. 2023. *Unidentified Anomalous Phenomena, Independent Study Team Report*. Washington: NASA.

Stump, E., 2003. *Aquinas*. London and New York: Routledge.

Thurston, H., 1952/2013. *The Physical Phenomena of Mysticism*. Guildford: White Crow Books.

Twining, N. F., 1947. *Twining Memorandum: Opinion Concerning "Flying Discs"*. Available at www.reddit.com/r/ufo/comments/eboml0/1947_twining_memo_ufos_are_real_and_not_fictitious/ (Accessed 20 July 2020).

Vallee, J. F., 1965. *Anatomy of a Phenomenon*. London: Tandem.

Vallee, J. F., 1969/2014. *Passport to Magonia from Folklore to Flying Saucers*. Brisbane: Daily Grail Publishing.

Vallee, J. F., 1975/2014. *Invisible College*. New York and San Antonio: Anomalist Books.

Vallee, J. F., 1988/2008. *Dimensions: A Case Book of Alien Contact*. New York and San Antonio: Anomalist Books.

Vallee, J. F., 1990. 'Five Arguments against the Extraterrestrial Origin of Unidentified Flying Objects'. *Journal of Scientific Exploration*, Vol 4, No 1, pp.105–117.

Vallee, J. F., 1990/2008. *Confrontations: A Scientists Search for Alien Contact*. New York and San Antonio: Anomalist Books.

Vallee, J. F., 1991/2008. *Revelations Alien Contact and Human Deception*. New York and San Antonio: Anomalist Books.

Weinstein, D. F., 2001. *A Catalog of Military, Airliner and Private Pilots sightings from 1916 to 2000*. Available at: Unidentified Aerial Phenomena – Eighty Years of Pilot Sightings: A Catalog of Military, Airliner, and Private Pilots sightings from 1916 to 2000 (squarespace.com) (Accessed 14 July 2023).

PART II
Ethics and Values

3

WE COME IN PEACE

Our Metaphysical Obligations to Aliens

Richard Playford

Introduction

The possibility of contact with extraterrestrial intelligences is simultaneously exciting and terrifying. It is exciting because of the ways it would transform our understanding of the universe, life, ourselves, and our place in the cosmos, whilst also allowing for scientific and political co-operation with these extraterrestrial intelligences in ways we can currently only imagine. At the same time, contact with extraterrestrials is not without its risks. A cursory glance over human history reveals that when "alien" human cultures come into contact, there is as high a risk of conflict as there is of co-operation. This raises a number of interesting questions: Is peace and co-operation with intelligent extraterrestrials desirable? What obligations would we have towards intelligent extraterrestrials? And should conflict occur, what norms should govern such a conflict, and what sort of a response would be required of us as individuals? In a sense, I want to offer a very simple answer to these questions: We would have the very same obligations towards these aliens as we do towards other human beings, and the same moral norms which govern our interactions with human beings would apply to our interactions with these extraterrestrials. I will not, per se, articulate any particular list of obligations we will have to these beings; rather I will simply argue that whatever obligations we have to other human beings, we will also have to extraterrestrial intelligences. I will do this by first explaining why we have obligations to each other. I will then explain why these same considerations hold for all intelligent animals, and thus why we have the very same obligations towards extraterrestrial intelligences as we do towards terrestrial intelligences. Finally, I will speculate as to whether we can hope that such extraterrestrial intelligences will also come to the same conclusions about their obligations to us.

Why Should I Care?

Why should we care about the well-being and rights of other people? Put another way, why do we have obligations towards others? I believe we have obligations towards other people because they possess two interrelated features. These are, first, that they have the capacity to

DOI: 10.4324/9781003440130-6

be harmed and benefited, and second, that they have the capacity to reflect on this capacity, to recognize it in others, and to communicate these facts to one another. Obviously, these claims require further explication, and so I shall discuss each in turn.

With regard to our capacity to be harmed and benefited, I take it that this capacity is uncontentious. We can clearly be physically harmed, through illness or injury, but as human beings, there are numerous other ways we can be harmed. For example, due to the sophistication of our psyches, we can also be harmed psychologically through the use of insults or deception. We also possess the ability to form life plans and goals, and these can be thwarted by chance or by the intentional actions of others. This too constitutes a form of harm. We are sociable creatures who partake in numerous social relationships. For example, we are members of small family groups, mid-sized local communities, and large, highly sophisticated state-level societies. As a result, we have an interest in maintaining these relationships and in the management of our reputation. As a result, we can be harmed through slander, gossip, invasions of privacy, infidelity, and so on.

The opposite side of the coin, of course, is the various ways we can be benefitted. As with harms, we can clearly be physically benefitted through the use of medicine when sick, or more generally through access to good quality food, exercise, and so on. Further, we can be psychologically benefitted by an appropriate work–life balance, good relationships, and so on. When we form life plans and goals, others can help us to achieve these, and, indeed, some of our life plans and goals can only be achieved with the help of others. Similarly, and on that note, when our families and communities flourish, we benefit from this as members of those families and communities.

We share this capacity to be harmed or benefited with all living creatures, but we also have certain capacities not shared by the vast majority of living beings. In this context, we can articulate this capacity as the capacity to be aware of our capacity to be harmed or benefited, to recognize the capacity to be harmed or benefited in others, to recognize the capacity in others to recognize the capacity to be harmed or benefited themselves, and to recognize in others the capacity to recognize the capacity to be harmed or benefited in others, and so on. Finally, alongside these high-level capacities which involve an understanding of ourselves as ourselves, others as others, and so on, we also possess the ability to communicate these insights to each other. Once I am aware that you and I are distinct, that we both have beliefs, but that those beliefs may differ, I can conceptualize the possibility of, and act to bring about, a change in your beliefs. The same also holds for our motivations and desires. We all have motivations and desires and these can be influenced by others in various ways.

This is all a very wordy way of saying that we have the ability to understand harms and benefits, both for ourselves and for others, to reflect on those harms and benefits and to communicate those reflections to others. Ultimately, this insight traces its origins back to Aristotle and is perhaps more eloquently expressed by him:

> That man is much more a political animal than any kind of bee or any herd animal is clear. For, as we assert, nature does nothing in vain; and man alone among the animals has speech …. Speech serves to reveal the advantageous and the harmful, and hence also the just and unjust. For it is peculiar to man as compared to the other animals that he alone has a perception of good and bad and just and unjust and other things of this sort; and partnership in these things is what makes a household and a city.
>
> *(Aristotle, Politics, 1.1253a)[1]*

Our capacity to communicate sophisticated ideas to each other is far greater than that of any other known species and, of course, communication requires groups of individuals to come together, and this is the beginnings of a community. It follows from this that human beings can engage in moral reasoning and can respond to the dictates of morality (however, these are conceived and whatever they may be).

This same insight has been articulated more recently in the work of John Rawls. His account of moral persons (or moral personality) whilst phrased a little differently seems to get at the same basic idea:

> Moral persons are distinguished by two features: first they are capable of having (and are assumed to have) a conception of their good (as expressed by a rational plan of life); and second they are capable of having (and are assumed to acquire) a sense of justice, a normally effective desire to apply and to act upon the principles of justice, at least to a certain minimum degree.
>
> *(Rawls, 1999, p. 442)*[2]

An important note of clarification is now needed to avoid an objection raised by Peter Singer (2011). He points out that such an account appears to fail to explain the moral worth of very young children and those with severe learning disabilities. As Singer explains,

> more serious is the objection that not all humans are moral persons, even in the most minimal sense. Infants and small children, along with human with profound intellectual disabilities, lack the required sense of justice. Shall we then say that all humans are equal, except for very young or intellectually disabled ones? This is certainly not what we ordinarily understand by the principles of equality.
>
> *(2011, p. 18)*

Singer then considers a potential response offered by Rawls, "Rawls deals with infants and children by including *potential* moral persons along with actual ones within the scope of the principle of equality" (Singer, 2011, p. 18).[3] The obvious problem with this response, at least when phrased like this, is that it appears to be an ad hoc solution, which also does little to account for the moral value of those with severe intellectual disabilities. As Singer explains,

> This is an ad hoc device, confessedly designed to square his theory with our ordinary moral intuitions, rather than something for which independent arguments can be produced. Moreover, although Rawls admits that those with irreparable intellectual disabilities "may present a difficulty", he offers no suggestions towards the solution of this difficulty.
>
> *(Singer, 2011, pp. 18–19)*[3]

Unfortunately, Singer's criticism misrepresents Rawls's position, and thus it fails as a criticism (both of Rawls and of my argument in this chapter). Rawls's does not believe that young infants are potential moral persons. Rather, he defines moral personhood in such a way so as to include very young infants. As Rawls explains "One should observe that moral personality is here defined as a potentiality that is ordinarily realized in due course. It is this potentiality which brings the claims of justice into play" (Rawls, 1999, p. 442).

The reason for defining moral personhood in this way is clear. As we go through life, we are not always able in the here and now to understand harms and benefits, both for ourselves and for others, to reflect on those harms and benefits, and to communicate those reflections to others. Drunkenness and unconsciousness, for example, can render us unable to exercise these abilities, but it does not follow from this that at these moments we are no longer moral persons. This is because we are still the sorts of creatures which can, in principle, understand harms and benefits, both for ourselves and for others, to reflect on those harms and benefits, and to communicate those reflections to others even though we cannot do so at that precise moment.

This observation is particularly important for the very young, who have not yet matured enough to exercise these abilities, and the disabled, who, due to their disability, are possibly forever incapable of exercising these abilities. The very young and the disabled are just as human as you and me, and human beings are the sorts of creatures who can understand harms and benefits, both for ourselves and for others, to reflect on those harms and benefits, and to communicate those reflections to others. Therefore, even though they cannot actually exercise those abilities in the here and now, they are still the sorts of beings who can understand harms and benefits, both for ourselves and for others, to reflect on those harms and benefits, and to communicate those reflections to others.

Rawls himself accepts this analysis for infants, but he rejects this analysis for those with severe intellectual disabilities. He holds that those with severe intellectual disabilities have lost the potentiality needed to count as moral persons. He then goes on to acknowledges (as Singer points out) that this poses a potential problem for him (Rawls, 1999, p. 446), since, presumably, those with severe intellectual disabilities are deserving of equal moral consideration as those without such disabilities.

However, I think the solution to Rawls's problem lies in his very writings. As Rawls points out, "When someone lacks the requisite potentiality either from birth or accident, this is regarded as a *defect or deprivation*" (Rawls, 1999, p. 443, italics added for emphasis). With this observation in mind, I think that the correct way to interpret those with severe intellectual disabilities is that they too are moral persons with the intrinsic potentiality to understand harms and benefits, both for ourselves and for others, to reflect on those harms and benefits, and to communicate those reflections to others that would be "*ordinarily realized in due course*" (Rawls, 1999, p. 442, italics added for emphasis), but which is prevented from being so realized by their disability. Whilst they may be unable to ever exercise this potentiality, it is still a part of them. If it were not, then it is unclear how we can understand their disability as a *defect or deprivation*. Neither a dog, for example, nor a human being with a severe intellectual disability can actually understand harms and benefits, both for themselves and for others, to reflect on those harms and benefits, and to communicate those reflections to others, but only one of them is suffering from a defect or deprivation. If we conceptualize moral personhood as a potentiality in this way, as a potentiality which can be permanently blocked by disability or damage, then we can justify our intuitively plausible claim that human beings with severe intellectual disabilities are deserving of equal moral consideration as those without such disabilities.

One objection that might be raised at this point is that this account of our obligations towards others fails to explain our duties towards (non-human) animals. After all, according to the account just given, animals lack moral personhood.[4] Therefore, if our obligations towards others stem from their status as moral persons, we cannot have obligations towards animals, and thus potentially we can do what we want to them, from a moral perspective, including things like torture and other such abhorrent acts. However, so the critic might argue this is absurd, and therefore there must be something wrong with this account of our moral obligations towards others.

This objection fails because whilst I have presented an account for why we have obligations towards other persons, there is nothing about this account to prevent there being other sources of moral norms and other reasons to treat animals well. As a result, whilst I have shown (at least one reason) why we have duties towards other human beings, this in no way rules out the possibility of duties towards, or in respect of, animals (or, indeed, the environment, nature, beauty, and so on). We might think, for example, that virtue demands we treat animals in a particular way, or that animals are deserving of certain sorts of treatment by merit of their resemblance to us in certain respects (but not others), and so on. As a result, whilst I am perhaps obliged to say that we have greater obligations towards our fellow human beings than we do towards animals, although even this is potentially open to debate, I am not obliged to say that we have no duties towards them and that they are not deserving of at least some moral consideration. This conclusion strikes me as highly plausible. Whilst it is plausible to claim that we have greater obligations towards our fellow human beings than we do towards animals, it would be highly counter-intuitive to argue that 'anything goes' when it comes to our treatment of them. The account I have just given of our obligations towards our fellow human beings strikes just such a balance!

Precisely, what obligations we have towards, or what moral norms there are with regard to our treatment of, animals goes beyond the scope of this chapter. Whether it simply rules out the permissibility of gratuitously torturing them, or whether it requires us to only eat free-range meat, or to become vegetarian or even vegan, would take us too far afield and thus I shall have to leave these questions unanswered. For now, however, we can confidently dismiss the concern that my account of our moral obligations towards one another would justify immoral acts against animals or render us unable to condemn them. For two discussions of how we might justify at least some moral norms with regard to our treatment of animals that would be compatible with the overall argumentation of this chapter, see Barad (1988) and chapter 3 of Oderberg (2000).

Looking after Our Own

We have now seen why we have obligations to our fellow human beings. However, it doesn't follow from this that we have equal obligations to all of our fellow human beings. The thesis that we have equal obligations to every human being is known as 'impartiality'. The partialist is one who holds that we are permitted or even obliged to treat certain others favourably, whereas the impartialist holds that we should treat everyone the same. I believe that partiality is the correct view to hold in this regard, and I believe a correct understanding of partiality is essential to understanding our potential obligations to intelligent extraterrestrials, so it is worth taking some time to explore this subject.

Whilst we possess some obligations to all human beings, we have greater obligations to those whose lives, and thus good, are more intimately tied to and intertwined with ours. In the real world, this means we have the greatest obligations to members of our own family and to very close friends, beyond that more distant friends and members of our local community, and beyond that still society at large, and then distant others. It is worth repeating that we have obligations to every single human being, but it does not follow from this that we owe a stranger halfway round the world the same level of consideration we do to our partners, parents, or children.

It should be clear why we have stronger obligations to our family members and close friends than to strangers. Our lives, goods, and well-being are intimately linked with these people, and as a result we are obliged to provide special care and consideration to our friends and family beyond the general beneficence and non-maleficence owed to strangers. This is part of what it means to be a friend or family member! Part of being a good friend or family member involves showing them preferential treatment under some, but certainly not all, circumstances.

At the same time, it is also worth adding that in the modern world all of us are interconnected. The actions and decisions that somebody makes at home can affect somebody on the other side of the planet and vice versa. We also live in an age with increasing amounts of travel, communication, and sharing amongst and between different, often quite distant, cultures. If we live in an even moderately large city, then the odds are that some of our neighbours may well have been born or have their origins in different countries and communities. This then connects our community with that community to at least some extent. As a result, modern technology, travel, and politics, as well as our shared history, bind all human beings together. In some sense, then we are all part of a global community, and this magnifies our obligations to distant others to a far greater extent than at any previous moment in human history.

Partiality doesn't strike me as contentious. If a father only has enough resources to, for example, educate one child, then, all other things being equal, it is reasonable for him to favour his son over a complete stranger. Similarly, if a mother only has enough time to nurture and care for one child, to hear their concerns and fears, and to offer them affection and comforting words, then, all other things being equal, it is reasonable for her to favour her daughter over a complete stranger, and so on. As John Cottingham points out,

> To choose to save one's own child from a burning building when an impartial consideration of the balance of general utility would favour rescuing someone else first, is *not* (as impartialists must claim) a perhaps understandable but nonetheless regrettable lapse from the highest moral standards; on the contrary, it is the morally correct course – it is precisely what a good parent *ought* to do. A parent who leaves his child to burn, on the grounds that the building contains someone else whose future contribution to the general welfare promises to be greater, is not a hero; he is (rightly) an object of moral contempt, a moral leper.
>
> *(1986, p. 357)*

Partiality, as Cottingham labels it, when applied appropriately is not a bad thing, indeed it is often morally required.

However, an objection that might be raised here would be to point out that sometimes partiality is a bad thing. As Cottingham points out, "To say without qualification that it is morally correct to favour one's own clearly will not do" (Cottingham, 1986, p. 358). Nepotism and favouritism can be immoral and damaging to society. A judge who allows a guilty criminal to go free simply because they are friends is not a just judge.[5] Cottingham also gives a number of other examples. He writes,

> A civil servant who in placing a contract shows favouritism to his friends or relations is (rightly) liable to be dismissed. Nor is the case of the public official the only type of case where partiality is impermissible. The personnel manager of a privately owned company may be censured for cronyism or nepotism even though he has sworn no oath of office to serve the public impartially.
>
> *(1986, p. 358)*

How are we to reconcile the claim that partiality is often a good thing with the observation that sometimes impartiality is required?

We can do this by remembering that the agents in the cases outlined earlier (the civil servant, judge, and personnel manager) are under either an explicit or implicit duty to be impartial. This is because they are representing organizations that need to be impartial rather than themselves as private individuals. The civil servant and the judge are acting on behalf of the state which should be impartial between its citizens. Further, the civil servant and the judge will most likely have taken an explicit oath or signed an explicit contract promising to be impartial when carrying out their duties. The personnel manager, whilst perhaps not having made an explicit commitment to be impartial, is acting on behalf and in the interests of a company which will need to be impartial in order to best serve its shareholders, employees, customers, and society at large. As a result, the personnel manager also has a duty to be impartial. As Cottingham explains,

> The situation in these various different cases, then, is that *either*, in what I shall call the "direct" case, (such as that of the public official) the agent is under a specific duty to be impartial; *or*, in the "indirect" case (such as that of the personnel manager) the agent is under a duty to perform an activity or job the requirements of which involve a duty to make non-biased assessments based on a range of objectively determined criteria. To take account of these complications, we need to redefine partialism as the thesis that *unless one is under a direct or indirect duty to be impartial, it is morally correct to favour one's own.*
>
> *(1986, p. 358)*

Another objection that might be raised here is that an ethical theory, which endorses partialism, is liable to endorse racism, xenophobia, and possibly sexism. After all, in some sense, two white Europeans have something in common that a person of African or Asian decent does not. Therefore, perhaps the two white Europeans should view each other as "one of their own" and favour them over people of other ethnic backgrounds. This would seem to be a troubling conclusion because this seems to be racist which is morally impermissible.

I shall put aside the question of whether or not patriotism is a virtue and instead focus solely on ethnicity for the sake of simplicity. Does partialism lead to or obligate racism? My answer is that partialism in no way necessitates or leads to racism because a person's ethnicity has no bearing on their moral worth or on our obligations towards them. As a result, all forms of racism, even if partialism is true, are impermissible. As we have seen, human beings have obligations towards one another because human beings have the ability to understand harms and benefits, both for themselves and for others, to reflect on those harms and benefits, and to communicate those reflections to one another. Ethnicity (and, indeed, sex, sexual orientation, and so on and so forth) are simply irrelevant since the colour of one's skin or geographical origins (for example) in no way affect one's ability to understand harms and benefits, both for themselves and for others, to reflect on those harms and benefits, and to communicate those reflections to one another. Further, there is nothing to stop us forming families, friendships, and communities with people from differing ethnic backgrounds. As a result, there is simply no reason to favour one's own race over other races. We can conclude, therefore, that partiality in no way leads to, or necessitates, racism.

Do *We* Come in Peace?

We now have the necessary intellectual resources needed to answer the question of this chapter: What are our obligations to intelligent extraterrestrials? The answer is that our obligations to intelligent extraterrestrials are exactly identical to our obligations to our fellow human beings.[6] Intelligent aliens, by merit of their intelligence, will be able to understand harms and benefits, both for themselves and for others, to reflect on those harms and benefits, and to communicate those reflections to one another. As a result, we will have obligations to them for the exact same reason we have obligations to each other. However, clearly this claim is in need of further clarification and explication, and so it is to these clarifications that I now turn.

First, I am not claiming that these intelligent aliens will be subject to, and will readily grasp, the same harms and benefits as us. Their physiology may be very different, rendering them immune to some of the harms that can befall us and vulnerable to some we are not, and likewise for benefits. Similarly, their psyches and societies may be structured differently to ours, and, thus, they may conceptualize psychological and sociological harms very differently to us. Despite this, however, their harms and benefits are just as much harms and benefits as our harms and benefits, and we owe it to them to consider these in our interactions with them and them in their interactions with us. Again, this shouldn't strike us as surprising. All of us have differing vulnerabilities, strengths, weaknesses, and so on. A good example of this is that of food allergies. Most people can eat peanuts without any ill effect, whereas for others they can be deadly. It does not follow from this that we can ignore the well-being of those with peanut allergies. Whilst all human beings have similar vulnerabilities, strengths, and weaknesses (none of us can walk through fire unharmed, for example), intelligent extraterrestrials may well have very "alien" vulnerabilities, strengths, and weaknesses, but this is a difference of degree rather than of kind. By this, I mean an alien vulnerability is still a vulnerability, an alien strength is still a strength, and an alien weakness is still a weakness. Of course, they may be different "kinds" of vulnerabilities to ours, but it remains the case that all vulnerabilities are, nonetheless, vulnerabilities (i.e., they are members of the kind "vulnerabilities"). In this sense, the difference between our vulnerabilities and alien vulnerabilities is one of degree, rather than one of kind. As a result, we owe them the same consideration we owe to our fellow human beings. The only thing that follows from this is the need for caution and prudence when interacting with intelligent aliens, should we ever find them.

Second, my observation that intelligent extraterrestrials may well have very different physiologies and psyches to us leads us to two interrelated worries. The first is that there may be physiological reasons why communicating with intelligent extraterrestrials is impossible. The second is that their psyches may be so different to ours that we cannot intelligently understand them. I shall discuss each in turn.

With regard to the possibility that there may be physiological reasons why communicating with intelligent extraterrestrials is impossible, the problem here needs to be correctly understood and its different components not conflated. One aspect of this worry is that because we cannot communicate with them, we cannot therefore have moral obligations towards them and them towards us. After all, the reason we have obligations to our fellow human beings is because we are all able to understand harms and benefits, both for ourselves and for others, to reflect on those harms and benefits, and to communicate those reflections to one another. If communication with these intelligent extraterrestrials is impossible, then it appears that the final aspect of this ability does not apply to these intelligent extraterrestrials.

We can obviously only speculate at this stage how likely, or unlikely, it is that we will be able to communicate effectively with intelligent extraterrestrials, and this speculation is probably best left to astrobiologists. However, let us imagine that we do encounter an intelligent alien species with whom genuine communication is impossible. Perhaps this is due to radically different ways of communicating between our species, for example, they communicate using very fine microwaves which even our most sensitive technology fails to pick up. Alternatively, perhaps there is some physical difference between us that prevents any real communication, maybe they live their whole lives in under a second or over a million years, and we simply move too fast or too slow compared to them in order to actually communicate, either way the precise details don't really matter. It seems that in cases like this we are incapable of truly communicating with these aliens. Therefore, one might argue that as a result of this we cannot have any obligations towards them, and thus we should treat them like we would wild animals.

From a practical perspective, there may well be a grain of truth in this. If we really are incapable of communicating in any way, then, in reality, the best action may well be to live and let live. We should leave them alone, and they should leave us alone because we are incapable of doing anything more than this. However, from a theoretical perspective, we would still have the same obligations towards them that we would towards other human beings. This is because they still share with us the ability to understand harms and benefits, both for themselves and for others, to reflect on those harms and benefits, and to communicate those reflections to one another. It just so happens that, due to an accident of biology, they are incapable of communicating those reflections to us in particular, and vice versa. This does nothing, however, to change their status as the sorts of creatures that can engage in moral reasoning, and thus be bound by the dictates of morality (however, these are conceived and whatever they may be). The differences between us and them (for example, in our modes of communication) are therefore contingent and accidental with regard to this. The fact that they communicate using microwaves, rather than sight and sound, is no more relevant to their moral status than the colour of a human being's skin.

Despite this, should we encounter such aliens, given genuine communication is impossible, it may well be the case that the best thing to do is to simply leave them alone and vice versa.

If we cannot ascertain what will harm and benefit them, what their intentions and goals are, and so on and so forth, and if there is nothing we can do to change this, the best course of action may well be to leave them alone and to live and let live.

At the same time, I am instinctively optimistic that physical barriers to communication can be overcome. Should we encounter intelligent extraterrestrials, then either we or they (or both of us) will have to possess some fairly sophisticated technology. Further, unless our understanding of physics and biology is radically wrong, there are only so many forces and processes at play in the world and only so many ways of interacting with them. Thus, it seems likely that our senses, and thus modes of communication, will overlap sufficiently so as to make communication, at least in principle, possible. For example, the sense of touch seems fairly fundamental and essential to survival, and it seems hard to imagine any even moderately sophisticated creature being unable to physically ascertain whether another object has come into contact with it. Using this shared sense of touch, at least in principle, a shared language could likely be created between us and any intelligent extraterrestrials, and thus communication would be theoretically possible.

David Oderberg humorously articulates this point using the example of a fantastical alien-like creature called "Glog" who

> has three heads full of liquid hydrogen, seventeen sensory organs, and twelve tentacles of varying lengths placed strategically around a spherical body of thixotropic clay. For nutrition it sucks in helium through a glass tube, expels xenon as waste, reproduces by shaking itself into liquid pools that reassemble into similar bodies, and gets about by magnetic levitation.
>
> *(2014, p. 219)*

He points out that even in the case of Glog, who is about as different from us physiologically as it is possible to currently imagine, there is no reason, in principle, why communication would be impossible. He writes:

> What reason of *principle* is there for saying that we could never communicate with each other? Suppose we could. Glog speaks via microwave pulses, so we use a microwave pulse detector to pick up the signals. Where we see trees, Glog sees vibrating atoms; so we tool up some appropriate scanning tunnelling microscopy for seeing what Glog sees. If we notice Glog acting as though in distress in a helium-deficient environment such as a chamber filled with xenon, we might think about adding some helium and taking out some xenon. And so on. Presumably Glog will try to thank us! Might it thank us by trying to exterminate us? If you think that a possibility, then you have not understood what it means to express gratitude. If Glog is rational, and it wants to thank us, it will not try to kill us. If it is rational, grateful and *evil* it might, but that is a different matter.
>
> *(Oderberg, 2014, pp. 219–220)*

Of course, all of this is pure speculation and rather fanciful speculation. As a result, the best advice is probably to cross this bridge when and if we come to it.

With regard to the concern that alien psyches may be so different to ours that we cannot intelligently understand them, I will not attempt to fully respond to this concern, but I will raise a few points for consideration. The first is that whether or not this is a possibility depends, in part, on how one conceptualizes reason and intelligence. If reason and rationality are objective and exist, in some way, "out there" and that when we reason we are trying to "follow" or "discover" these dictates of reason, then it follows that an intelligent alien species will be at least attempting to do the same thing. Therefore, if we reason correctly, we will find each other mutually intelligible.

Alternatively, we may think that reason and rationality are, so to speak, "foisted upon us" by our brains (or minds, or bodies, or whatever). If this is the case, then alien "reason" may well be completely different to ours and thus mutually unintelligible.

To weigh in on this debate would lead us into a number of highly thorny issues that go beyond the scope of this chapter and the space available; however, I will make two observations. The first is that the success of mathematics, both in and of itself, and as a tool for understanding the physical world, inclines me to adopt the first model of reason and rationality. If reason and rationality are simply foisted on us by our brains, for example, and shaped entirely by evolutionary forces, for example, then it is entirely unclear to me why the human mind, and mathematics, is so successful when it comes to understanding deep truths about the physical universe. I acknowledge that this observation is in no way decisive, nor is it meant to be, but it does make me favour the first option. Second, if one chooses to adopt the second option, which allows for the possibility that alien "reason" may well be completely different to ours, then I would point out that by definition we cannot "reason" about it in any objective sense of the word. We can only reason using our "reason", and that is exactly what I am doing. Anything beyond this, if done by a human being, by definition, is unreasonable since it goes beyond the limits of our reason. As a result, I conclude that we should assume that they will reason like us. No doubt aliens will have their own "take" on various problems, and their own intellectual strengths and weaknesses, interests, and blind spots, quite possibly shaped by their physiology and environment, but despite this, I think we should assume that if we and they reason correctly (and if we find a way to communicate those thoughts to one another), we will find each other mutually intelligible. If we cannot understand them, I suggest we simply reason "harder"![7]

The third point needed clarification concerns partiality. I have already endorsed partiality and have argued that it is permitted to prioritize "our own" over strangers. This might lead some to conclude that we should therefore always prioritize our fellow human beings over intelligent extraterrestrials; however, this conclusion is incorrect for a number of reasons. First, whilst it is permitted to prioritize our family members, for example, over strangers, this is not an absolute claim. If a stranger's need is far greater than a family member's, then it is perfectly appropriate under those circumstances to prioritize the stranger. Whilst it is appropriate to feed your family first, it is inappropriate to feed them seconds whilst a stranger starves.[8] As a result, even if human beings, by merit of being human, are automatically "our own", it does not follow that we should *always* prioritize them over intelligent extraterrestrials.

Second, and just as importantly, it is simply not the case that human beings will automatically always be "our own" over and above intelligent aliens, at least not once we have developed bonds of trust and friendship with them. As we have seen, what matters from the moral perspective is the ability to understand harms and benefits, both for themselves and for others, to reflect on those harms and benefits, and to communicate those reflections to one another. Everything else is simply irrelevant. Thus, in the same way that race is no barrier to forming a family, friendship or community with someone of a different race, neither is alien physiology! As a result, intelligent extraterrestrials could form friendships and communities (and families?) with us, and once they had done that, we would owe them all the same partiality we owe to our human friends and community members (and perhaps even family members).

The overall conclusion we should take from this section is that should we encounter intelligent extraterrestrials, they will be deserving of the same moral treatment due to us and our fellow human beings.

Do *They* Come in Peace?

So far, I have argued that the reason we have obligations to our fellow human beings is because we have the ability to understand harms and benefits, both for ourselves and for others, to reflect on those harms and benefits, and to communicate those reflections to others. I then examined partiality and defended the claim that it is sometimes permissible or even obligatory, but that it cannot be used to justify racism, xenophobia, and so on. I then turned to aliens and applied these insights to intelligent extraterrestrials. I showed that it follows from my analysis of the origins of our obligations to our fellow human beings that we will have the same general obligations to intelligent extraterrestrials, whilst acknowledging that practical realities may oblige us to treat them differently in the specifics. I then considered partiality and acknowledged that partiality *could* sometimes make us favour our fellow human beings over intelligent aliens given we have bonds of trust and friendship with (at least some of) our fellow human beings, but that this should not be overstated and that the opposite may be the case once we have developed bonds of trust and friendship with an alien species.

To conclude this chapter, I now want to consider whether or not we can hope that intelligent aliens, should they find us, will come in peace. Will they recognize our intrinsic moral worth? Will they respect our rights and treat us with beneficence and non-maleficence?

Of course, we can only speculate at this stage. Despite this, again, I am instinctively optimistic. The only rational species we can draw inferences from is ourselves, and whilst human history is full of exploitation and cruelty, it is also full of kindness, co-operation, charity, and so on. Additionally, in certain respects at least, I think we can see evidence of moral growth. Racism and slavery were once commonplace, and whilst these sadly still exist in the modern world, we are gradually coming to see the wrongness of these practices and beliefs.

To these observations we can add that if, as I have argued, our morality is tied to our rationality, it follows that other intelligent species will likewise have a moral sense and will engage in moral reasoning. Further, if an alien species were to contact us, they would have to be technologically advanced and thus intelligent. From this, we can conclude that they too will engage in moral reasoning and will hopefully have come to similar observations about their obligations to us as I have argued we have to them in this chapter. Therefore, we can reasonably hope that should alien species contact us, they will come in peace.

Who knows? Perhaps, on an alien planet somewhere amongst the stars, a very exotic-looking creature is reading a strange-looking book written in an alien language arguing for the very same conclusions as this chapter does here on Earth. That strikes me as an encouraging possibility!

Notes

1 See also St Thomas Aquinas's *Commentary on Aristotle's Politics* (Aquinas, *Comm. Nic. Ethics*, I.1.21).
2 See also Rawls (1999) p. 11 and p. 17, and Rawls (1980) p. 525.
3 Singer does not provide a reference for Rawls here, and so I cannot be certain to which of Rawls's writings he is referring. However, I suspect it is Rawls (1999), section 77, which discusses these issues at length.
4 Some might argue that cetaceans (dolphins, whales, and so on) and great apes (chimps, gorillas, and so on) are, or at least could be, moral persons. There is insufficient space here to examine this possibility and it goes beyond the scope of this chapter, so we will simply have to put the moral status of cetaceans and great apes aside for now. If they do count as moral persons, then that does little to undermine the overall thesis of this chapter, and if they don't, then they will simply have the same status as other "lower" (non-human) animals.
5 In countries like the UK, of course, judges would not be allowed to preside over a case involving friends or relatives, as there would be an obvious conflict of interest. As a result, a situation like this, in theory, would never happen. In practice, of course, conflicts of interests do emerge in our judicial systems despite our best efforts to avoid them. Nevertheless, my point still stands and indeed is supported by the observation that we have practices like this. There are times when impartiality is required, even if there is nothing wrong with partiality in other contexts.
6 According to Aristotle's definition of a human being as a "rational animal", intelligent aliens would also count as "human beings". At the same time, not everyone subscribes to Aristotle's definition of humanity or to his "species concept". As a result, I have used, and will continue to use, the word "human being" as a synonym for *Homo sapiens* in an attempt to avoid controversy. I take it that, regardless of your definition of human being, and regardless of your view of the "species problem", intelligent extraterrestrials would not be, and should not be, labelled *Homo sapiens*.
7 For a lengthier discussion of at least certain aspects of this issue, see Evnine (2001).
8 This shouldn't be taken as a precise moral claim, nor concrete piece of advice, for the real world. I'm simply using it as a possible illustration of my broader point.

References

Aquinas, Thomas (2007) *Commentary on Aristotle's Politics*, translated by Richard J. Regan, Indianapolis: Hackett Publishing Company.
Aristotle (1998) *Politics*, translated by C. D. C. Reeve, Indianapolis: Hackett Publishing Company.
Barad, Judith. (1988) 'Aquinas' Inconsistency on the Nature and Treatment of Animals', *Between the Species*, Vol. 4, No. 2, pp. 102–111.
Cottingham, John. (1986) 'Partialism, Favouritism and Morality', *The Philosophical Quarterly*, Vol. 36, No. 144, pp. 357–373.
Evnine, Simon J. (2001) 'The Universality of Logic: On the Connection between Rationality and Logical Ability', *Mind*, Vol. 110, No. 438, pp. 335–367.
Oderberg, D. S. (2000) *Applied Ethics: A Non-Consequentialist Approach*, Oxford: Blackwell.
Oderberg, D. S. (2014) 'Could There Be a Superhuman Species?', *The Southern Journal of Philosophy*, Vol. 52, No. 2, pp. 206–226.
Rawls, John (1980) 'Kantian Constructivism in Moral Theory', *The Journal of Philosophy*, Vol. 77, No. 9, pp. 515–572.
Rawls, John (1999) *A Theory of Justice*, revised edition, Cambridge: Harvard University Press.
Singer, Peter (2011) *Practical Ethics*, 3rd edition, Cambridge: Cambridge University Press.

4

HEGELIAN ALIENS

A Call for Cosmic Completion

Lewis Howeth

Part 1: What Is the Good?

Aristotle argued, "every craft and every line of inquiry, and likewise every action and decision, seems to seek some good; that is why some people were right to describe the good as what everything seeks" (Aristotle, 1999, p. 1). To paraphrase, all goals or "ends" presuppose a good, a reason to be sought after, as Aristotle presumes that nature is always purposeful and logically directed, even considering the most brutish animals.

Aristotle argues that particular desires reveal an underlying inclination towards goodness, such that all particular goods are subordinate to something good in itself. Aristotle believes, we enact our desires often not for their own sake but for the sake of something higher we wish to achieve. An example might be the good of going to the gym; it may have practical goods such as the enjoyment of lifting weights or a runner's high; however, those goods are only properly realized as goods within a whole life, which might also include beauty and health, amongst other possible goods. Taken to its logical end, Aristotle argues that all goods share the same overall aim. We can reach the truth of this more excellent end by reflectively regressing our desires to what is supremely good and desirable for itself alone; so, the good is that which is most desirable. Aristotle's equivalence of goodness and that which is desirable allows for a concrete referent for the subject's moral inquiry. This provides us with the means to bridge the gap between metaphysics and ethics; in our case, it opens the door to the metaphysics of desire.

In order to gain a solid understanding of his metaphysics of desire, we must first outline his general metaphysics briefly before delving into his metaethics. Aristotle argues that we can know a being through its substance; a substance is *that* which remains after all characteristics of an object have been alienated. Characteristics which are alienable from an object whilst leaving the object with the same identity are called "accidental". For example, a cat could be of any colour, hairless, of any particular breed, and still be a cat. Likewise, if you remove any of these traits, the essence of the cat persists; Aristotle calls these traits accidents, qualities which mark the particularity of a being but aren't required. The substance of the cat is then what cannot be removed from any cat without rendering

DOI: 10.4324/9781003440130-7

it a different identity, for example, it's being a mammal which could not be removed from the general nature of cats, or any cat in particular, without rendering the being something other than a cat.

In order to properly set the foundations of our development into Hegelianism, it's important to note that a substance is not the same as essence. An essence, like a substance, is the irreducible qualities of an object; however, unlike a substance, it may not exist in reality; for example, a unicorn would have the essence of being a horned horse, but that does not make it real and therefore substantial; so, an essence is purely formal, lacking matter. As Aristotle says, "By the substance without matter I mean the essence" (2016, p. 113). The substance, then, is the natural species of a being which individual members can be said to exhibit, and the essence is the substance considered abstractly.

Each substance can be further understood in relation to four causal components, which Aristotle attests exhaust the truth of a being when combined. These are the material, formal, efficient, and final causes. The material cause is that which holds the form or essence in reality, not necessarily matter as we perceive it today. The formal cause is the essence of a being, as mentioned earlier. The efficient cause is that which led to the existence of a being within a causal chain (e.g., a carpenter to a table), and the final cause is the end to which a being is changing towards (e.g., a seed is changing into a tree). Armed with this very brief introduction to Aristotelian metaphysics, we can now delve deeper into his metaphysics of desire.

Aristotle understands desire as a drive, a process which moves a subject towards or away from sensed objects, and as he phrases it:

All animals have at least one kind of perception, touch. And that to which perception belongs, to this belongs also both pleasure and pain, as well as both the pleasurable and the painful; and to those things to which these belong also belongs appetite, since appetite is a desire for what is pleasurable.

(2016, p. 27)

This drive towards what is pleasant, he notices, has metaphysical implications; the movement itself implies a change. Change occurs when something moves from potentiality to actuality. Potentiality refers to what an object can become. Actuality, on the other hand, is what an object is. An example would be a seed and a tree; to define a seed is to take its present existence with it's potential to grow into a tree; the tree is likewise to be understood in its relation to producing seeds and its historicity, including being a sapling and a seed. The whole process is understood as the progenesis of a species, linking each particular stage of each specific plant with previous stages and stages yet to happen. Thus, Aristotle is arguing its substance includes these considerations which unite the continuity of each particular being within its substance.

Having now completed this brief introduction to Aristotle's metaphysics, we can hone in on the specific metaphysics of desire. As noted earlier, Aristotle claims that desire is an "appetition of what is pleasant", which implies an acknowledgment of a pre-existing relationship between the subject and possible goods. This relationship further implies that a subject appreciates certain ends and so has the capacity to be his own efficient cause in acting towards those ends. Aristotle acknowledges this difference between plants and animals. Both plants and animals share what he calls the nutritive drive, which is the means of converting their environment into the means of their flourishing, such as they both seek out and utilize water. Plants, however, lack the particular sort of self-directed movement or behaviour which animals exhibit. As a result, both the nutritive and the appetitive imply a relation to a final cause, which marks the ideal end goal of the species; the plant grows towards maturity by seeking nutrition, as does the animal. However, the animal can consciously modify itself and its world to construct the conditions for its flourishing. This capacity to consciously strive for flourishing marks the plant/animal distinction within Aristotle. The plant is entirely unaware of its end, so it is moved; the animal is aware of specific goods and moves. These distinctions also apply to alien life; we might find that a being actively seeks some state and avoids others; or we might find another which is alive but doesn't deliberate upon its behaviour to aid itself. The way in which a being functions towards its end marks what Aristotle describes as its soul, as Aristotle says, "If an eye were an animal, its soul would be sight, since this would be the substance of the eye corresponding to the account" (Aristotle, 2016, p. 23). This soul is, therefore, the directed animating force within a creature, determined by its species. The rational soul is able to choose its end upon reflection marking another distinction. This is why Aristotle sees man as the perfected animal, since we are able to correct issues within our soul and so put ourselves on course for our flourishing. All beings, however, are aiming towards an end, plants passively, animals actively, and humans rationally.

This offers the first glimpse into moral categorization relative to the substance of a specific alien species. We must also be aware of potential differences in forms of life; animality might look radically different for extraterrestrials who have undergone evolutionary progress in radically differing conditions. An example of this phenomena is outlined in the science fiction show *Star Trek*, in the episode "Home Soil", which pictures a creature made entirely out of minerals, forming crystalline structure which allows for biochemical and energy movements that resemble a brain. At first, they are mistakenly categorized as non-biological since these entities do not move like most animals but have consciousness. They live like filter feeders, and so we could imagine a crystalline entity might need to resolve internal, rather than external, problems for their survival and flourishing, such as best-possible structures and growth patterns, etc. What this demonstrates is our potential inductive bias when interacting with alien life. Such a bias can be confronted directly and challenged but likely never eliminated entirely.

We have now shown that given the Aristotelian account, there is an ontological relationship that all beings have to themselves, which indicates their good and that this good is the goal of all beings. However, beings do not exist isolated from others and must come about for some reason other than themselves unless they are *causa sui* and so contain the reason for their own existence, but we cannot make such a claim about any natural (i.e., created) being. Aristotle sees this as part of a whole, a larger process of nature or existence itself moving towards a natural good in which each substance, and its being, is caused by and subsequently drawn towards its end; he calls this the prime mover. Each substance then has its place in a natural order or law-like structure which determines it relative to itself and others, a process which explains its existence and one which matches nicely onto the process of natural selection. All life, whether extraterrestrial or not, is governed by this process, but does this process aim towards a single good or are their multiple goods which the being achieves throughout this process? We are now burdened with the age-old question of whether the good is one or many.

Before we jump head-first into another metaphysical rabbit hole, let us talk about the significance of answering the question. At face value, there are two possible solutions. The first is a universal good that all things share; the second is that multiple goods are individual and unique to all beings. Once more, this solution will be tied to, ontologically speaking, the existence of a substance, whether a substance is one or many. Let us take the answer to make that which is good to be specific to a particular being, such as a single human. Our consideration of good might be purely egoistic, perhaps in the frame of Hobbes (1998), with all the ethical and political ramifications implied. Likewise, if we say the "Good" is universal and distinct from the particularity of a being, as seen in the philosophy of Plato (2013) or Kant (2018), with its specific ramifications.

In the egoistic scenario, we might make a peace treaty with aliens because the result would be heavy casualties within an interstellar war. However, changing the balance of power might lead us to exploit them for our good. We also may have to rely upon a social contract, as seen in Hobbes, to gain political stability; if one cannot be made, war may be the only and even natural state of affairs. Thus, it might be rational to conclude that it would be better for all parties if we did not meet. This is the conclusion reached by those who propose the dark forest hypothesis, which answers the Fermi paradox by attributing a fear of discovery as a motivation for hiddenness.

Alternatively, the universalist may seek peace or war on rational grounds alone, citing a moral imperative towards a general good. We may also be obligated to seek out new life and civilizations to offer aid in order that no need goes unrecognized. Now that we understand the stakes, let's engage with the question.

Part 2: Is the Good One or Many?

The problem of the one and the many is a metaphysical problem that has been iterated in many forms since Aristotle himself. The problem is a metaphysical issue which seeks to outline the defined unity versus diversity of identity or, more colloquially, what establishes the relation between the whole and its parts.

David Lewis (1941–2001) expresses the problem as the boundary of a cloud. From the ground, we perceive a sharp boundary; however, the cloud is just a collection or aggregate of water droplets These droplets are denser in the centre and less dense on the outskirts until eventually there are very few. Therefore the question is, what marks the boundary and what constitutes a cloud? Do we have numerous clouds and only perceive one, or is there no cloud at all and only droplets? He remarks,

> Since they have equal claim, how can we say that the cloud is one of these aggregates rather than another? But if all of them count as clouds, then we have many clouds rather than one. And if none of them count, each one being ruled out because of the competition from the others, then we have no cloud. How is it, then, that we have just one cloud? And yet we do.
>
> *(Lewis, 1993, p. 23)*

The problem is metaphysical. What allows us to say that there is a single unified entity rather than a plurality? Likewise, what gives us the means of saying that there is an identity if the plurality falls into absolute particularity? We can reduce the cloud to water, the water to atoms, the atoms to quarks, but the problem continues since we can always postulate a smaller or more perfect unit. We are seeking a boundary, or boundaries, which can explain simultaneously the existence of the cloud, the droplets, the water, or any structure, however big or small, and yet it is not clear where such a boundary, or boundaries, should be located or how it, or they, should be conceptualized. It is clear that every unit or particular, no matter how small, requires some form of unifying principle which makes it a whole rather than disjointed parts, as is the case of the Aristotelian notion of substance. The issue is as follows: Aristotle conceives of being and goodness as interchangeable; since potentiality is a movement towards its completion or perfection, this movement from potential to actual is the movement from non-being to being and so from less to more good. Aquinas states it as follows: "For all desire their own perfection. But everything is perfect so far as it is actual" (Aquinas, 1911, p. 53). This means that change, including actions, are directed from a lack of perfection towards perfection. This very well may place the end (i.e., good) of each creature beyond the bounds of its species and instead in nature or existence itself. Hence, the problem of the one and the many for creatures directly mirrors and interrelates with our question on whether the "Good" is one or many. Wherever we draw the line for being, we also draw one for goodness as well.

Goodness for Aristotle is a being achieving its end, like the seed actualizing it's potential to become the tree. The movement is one from potentiality to actuality in which latent tendencies of the seed are realized through the biological progression towards maturity. The tree has fully actualized its latent potential. For this reason, Aquinas argues that goodness and being are convertible, since the ascertainment of any good can be seen as the movement from a potentiality to an actuality, or to put it another way, a being in order to change must lack something. Thus, a being which is perfectly good would be perfectly actual since it would lack nothing and therefore never change.

So are finite goods really good? Are more prolonged goods of an entire species really good? Are the goods of animals, plants, or life itself real goods, or is there only one good? Or in their competition do they annihilate any and all bounds of goodness? Aquinas answers this for us.

Aquinas argues that what is good is neither one *or* many but one *and* many. Aquinas acknowledges both possibilities, pointing out the contention when we first engage with the question. He begins by first outlining the Platonic concept of the good as the one, which he describes as absolute being or oneness and argues that "and by participation of these everything was called being or one; and what was thus per se being and one, he said was the highest good (summum bonum)" (Aquinas, 1911, p. 69). This is because in the Platonic case, that being which unites all existence is the only true good, and all other goods, for example, human goods are imperfect goods, leading him to believe that the only real goods were achievable through rational contemplation and valid only in how they participate in this one good. In attempting to consider everything, we abandon that particularity which marks the defined limits of a being, which makes a being one and distinct from other beings; in the case of the cloud, we would be losing it to the sky. In other words, there would be no substantial good for humans or aliens, only the good of God or being itself or none at all. On the other hand, if we take good to be particular, there is nothing left to talk about since we cannot give the real concrete conditions which mark any united form of goodness. However, Aristotle remedied this problem as Aquinas says there is a single perfectly actual being which is called Good, but through our participation in that being as limitations, we have our own goods. These goods are similar but not identical; they are made possible by their participation in this one good; for example, human goods are a limited instantiation of this one good; however, they are real and distinct existences and so real and distinct goods. They are good in themselves in that they are actual, and they are good in their participation within this one good; and so according to Aquinas, "Thus there is one goodness belonging to all, and also many kinds of goodness" (1911, p. 70).

Aquinas demonstrates that this problem is resolved by acknowledging that we can conceive of distinct species, such as humanity, but only as much as it is a limited part of existence. This makes the existence of species self-contained, and so, by definition, makes their goods their own. We can thus be assured that there are individual and universal goods, but that these individual goods are limited and must partake in the one absolute good which acts as the ultimate end of all beings. This, for Aquinas, leads to ethics being expressed most clearly as love, such as found where Jesus commands us to first love our God with all our heart, soul, and mind, and then similarly that we should love our neighbour as we love ourselves (Mathew 22:37–40). This is because love is the first act of the will, as Aquinas states:

> Love, however, regards good universally, whether possessed or not. Hence love is naturally the first act of the will and appetite; for which reason all the other appetitive motions presuppose love, as their root and origin.
>
> *(1911, p. 284)*

Love is directed towards that which is universally good, so all particular goods and particular desires are spawned from this first relation. In the Thomistic view, this would be the love for God. In the Platonic view, it would be the love for the good. This is why we could say that Jesus' words on the similitude of loving thy neighbour and loving God correspond with Aquinas' view, as through this initial universal love we gain specific loves and specific appetites. In other words, we must first love a universality of goodness, and then we can limit ourselves to more specific and qualified goods, such as human or alien goods. Thus, to obtain our goods, we are bound to consider the goods of the other necessarily.

Metaphysically speaking, if any substance is an imperfect instantiation of a universal substance, then that substance and its associated goods are made possible through its participation in the universal good. Since the universal good necessarily contains all other goods, it implies that our happiness must be inclusive of all goods. Since in desiring the universal, we gain the particular but not only our own particular goods but all particular goods. Thus, in making the first desire universal, Aquinas has indicated a moral obligation to all ends is the foundation of the virtue of charity and mutual association, thus marking friendship as "concomitant with perfect Happiness" (1914, p. 68). Hence, in order to achieve our own goods, we are forced to consider the goods of others. Aquinas does make a caveat, however; this friendship must be reciprocated and so it is only possible towards rational animals since it represents what he and Augustine call our "other self". Non-rational animals, however, since they can't reciprocate cannot form this bond. In the case of aliens, this seems to make us morally obliged to care for rational extraterrestrials within the confines of love.

With respect to extraterrestrials we can safely say we share a single unified good yet have separate and unique goods. We are obligated towards caring about their good and vice versa, but in terms of how we should structure our normative ethics and specific moral obligations, we are here left with two issues. The first is whether or not Aquinas is correct to say that we are only obligated to those who can reciprocate our care, as previously explained? The second is how do we judge what is in the interests of the other? The former question will be more easily addressed after we resolve the latter. We must ask the question: How do we love an alien?

Part 3: How Do We Love an Alien?

Let us remember that alien etymologically comes from the Latin *alius*, meaning other. The etymology clarifies the difficulty, how do we love that which is other? Love is, for Aquinas, as stated earlier, the appetite for what is good in another. However, for us to truly love another, we must be able to apprehend their good. Aquinas for this reason describes love as the cause of "indwelling" or mutual recognition of each other's goods or most specifically, each other's souls. It is therefore our obligation to train ourselves to perceive the proper function of a being so as to identify the good they are to aim for, since the soul of a being is the end or function of a being, as referenced from Aristotle earlier. For Aquinas, we are driven to go beyond a superficial apprehension of each other and instead seek a more intimate knowledge of each other. This should be reasonably intuitive. Otherwise, our acts of love will be misplaced and misguided or, as Alan Watts (1972) old joke goes, "Kindly let me help you, or you will drown, said the monkey, putting the fish safely up a tree". We may very well be acting with good intentions, but we will fall short of love as loving action is not merely goodwill but apprehension of the soul of the other. So, then we have realized that to love is to desire the good of the other, that this desire includes an epistemic duty; without as much, we can never hope of being loving. Thus, we need a way of obtaining a means of demonstrating our love for our alien-self.

Hegel offers us a systemic and Aristotelian way of how this process of mutual indwelling functions. Hegel, although distinct in many ways from Aquinas and Aristotle, most notably through his adoption of Kantian and post-Kantian German thought, is still very much in line with the Christian–Aristotelian tradition. Hegel was a Protestant thinker who adopted much of Aristotle's thought, especially on the topic of mind and soul. He even went so far as to claim that Aristotle's works on the soul are the foundation of philosophy of mind such that "The essential aim of a philosophy of mind can only be to introduce the concept again into the knowledge of mind, and so also to disclose once more the sense of those Aristotelian books" (2010, pp. 4–5).

Hegel based his system of logic and mind on Aristotle, as well as the Scholastics like Aquinas and Anselm. He agreed with many of the Aristotelian positions thus made, with the primary difference that he did not believe we had unmediated access to reality. Instead, Hegel argued that reality, as we know it, is best understood as Mind (Geist) in view of the fact that it is reducible to abstract logical principles. Hegel adopts this position as he affirms that Kant's critical philosophy has torn empiricism asunder, making the mind an integral component of reality. Kant (2000) demonstrated in his *Critique of Pure Reason* that apperception structured reality by applying categories such as space and time, in which we turn sensed reality into an appearance and experience of reality. Essentially, Kant argues that we structure reality through logical relations which do not necessarily exist externally to the agent. Hegel combats this view, arguing that although Kant is correct to see the logical operations as necessary functions of experience and relate these functions to a mind, he incorrectly attributes these functions to a subjective mind instead of reality itself. This is because in order for us to give an account of subjectivity, there must be a non-subjective logical structure we can reference. Hegel is then rejecting the Kantian concept of a thing-in-itself or noumena (unconditioned reality), instead positing that the categories of the understanding are an objective and subjective component within reality, so our understanding attempts to cohere the subjective categories with real substantial identities. This places Hegel firmly as a natural law theorist. However, it is a development away from the naive realism of pre-Kantian thought.

Hegel describes himself as an absolute idealist. He views reality as a unity of diametrically opposed operations which combine to form distinct ideas and ultimately a universal mind. When taken to their totality and united, they encompass all reality as a unified immaterial idea. This idea is governed by an intrinsic logic or logos, a real set of categories which govern what was, is, and can be, and also what can be judged or thought. The logic of this structure is subdivided into three:

1 The Doctrine of Being.
2 The Doctrine of Essence.
3 The Doctrine of Notion and Idea.

That is, the Theory of Thought in:

1 Its immediacy, the notion implicit, and in germ.
2 Its reflection and mediation, the being-for-self, and show of the notion.
3 Its return into self, and its developed abiding by itself – the notion in and for itself (Hegel, 1991, p.133).

This trinary view Hegel refers to as the unity of opposites, a concept he owes to Heraclitus (1987) and Plato. The unity of opposites begins from purely analytic relations, namely identity and difference, which he believes marks the first relation, since identity is self-relation and difference is other-relation. These are diametrically opposed yet mutually dependent operations. This unity in opposition can be demonstrated in their attempt to be taken separately, for example, we imply the operation of difference to identity when we assert that identity is different from difference; likewise, we refer to the difference as a distinct identity. This forms the basis of numerous contraries and the basis for the problem of the one and the many, as discussed earlier. Hegel, like Aristotle, sees the problem of the one and the many resolved into their unity, arguing that the logical operations, which derive their meaning, are mutually codependent, and that only when taken together in appropriate proportions do we have the Truth. However, Hegel (1977) explains Aristotle's notion of change through what he calls Dialectic. The Dialectic is the logical back and forth between these logical extremes played out in reality. Hegel believes dialectic is the foundation of all physical mechanisms, the development of chemical identity, the development of life, and the three types of soul Aristotle observed.

What is essential to understand in Hegel is that the teleological nature of reality is causing this dialectical movement which subsumes everything, including Ethical Life (Siltlichkeit). The end or telos is a perfectly united identity which he titles *absolute spirit*, and this has a retro-causative role which draws everything towards it. Akin to Aristotle, this being is perfectly actual and so perfectly good, and so our attempts to know particular and individual goods drive us towards this absolute Good. The process being dialectical can be thought of as a striving for stability and harmony, all identities, including those of ethical agents, are either reacting or responding to the world seeking to exhaust their potential. All moving towards something greater but pushing and bumping into each other along the way, giving and taking from each other in an attempt to achieve this end. Hegel perceives this process as a constant overshooting by contingent beings actualizing themselves, leading to one-sided relations which are unstable and can only be resolved in an opposing movement. This makes the process imminent, ever developing, and linear. Hence, Hegel famously perceives history as progressive.

The opposing movement of history is due to the logical operation of negation and sublation or *the negation of negation* (Aufhebung). From Hegel's perspective, Being and Essence are continuously negating one another since they are contraries and together form the basis of what he calls the Idea. Being and Essence negate one another until a perfect unity or idea is reached, and the negative falls away to reveal the unity of absolute truth. This perfect idea is an absolutely universal self which Hegel perceives as God. Therefore, we should perceive the end as a thinking being thinking of itself eternally, a being who experiences perfect unity of their being and essence. This is a traditionally Christian conception of God which we can see in the work of Aquinas; since God is purely simple, he is the perfect unity of existence and essence. This is why Hegel believes subjectivity is inevitable since the end is a perfectly self-conscious being, and so our rationality develops on nature's path of self-realization, thus, producing the nutritive, appetitive, and then the rational soul. This is identical to the Aristotelian format, with one caveat. Rationality is formed through negation of animality. Rationality stands above the instinctual drives of the animal operating as a counterbalance capable of vetoing any drive or action, marking the basis of our free will. The rationality is not truly free, however, until it negates itself to reveal that it's a rational animal; rejecting all desire and drive leads only to meaninglessness and its own destruction. Thus, it is only a negation of our brutish world view, not our animality. This process of revelation is twofold; first, through direct feedback with reality, such as frustration, confusion, indecision, suffering, and general unhappiness (practical reason). Second, through our engagements with others (theoretical reasoning).

Like Aquinas, Hegel sees mutual recognition as necessary for us to conceive of our true ends. This is because the other has a logical role, namely negation. The other negates our world view and our desires and makes available for the first time an outsider's perspective of ourselves. For example, a drug addict might be unaware of their addiction. They may be dependent but believe that, despite their unhappiness, the drug is a source of pleasure, relief, and even well-being, that is, until someone comes along and bursts their bubble. We are revealed to ourselves through the other in which we gain a whole new perspective. This is because the others' subjectivity negates our own and we negate theirs. This means that we go from practical reasoners to theoretical reasoners, capable of perceiving abstractly and separating ourselves from our desires. This allows us to conceive of a new identity for ourselves, and so we can now conceive of our own essence. However, the struggle for recognition is not only proposing an essence but managing to create the correct essence, this is why both the theoretical and practical reason are necessary and when taken together form the basis of morality.

Moreover, this essence is the product of a process which subsumes the reason of another, making it a double negation. This means that both parties negate their private conceptions and instead produce together a new shared identity. This shared identity is the basis of ethics and marks the beginning of man's political life, as such it is corrupted by our vices and ignorance. We produce an unfair, unjust, and incomplete identity, one which places the interests of individuals and groups above others inherently. Hegel describes this as the Lord and Bondsman relationship, in which one party acts as the lord reaping the rewards of the relationship, while the other is the worker fulfilling the interests of the lord. However, since we gain feedback from our lived experiences, the worker comes to realize his dissatisfaction and opposes the lord's rule. This is the beginning of a constant and prolonged struggle for mutual recognition, as Hegel describes this as the notion of recognition:

> The Notion of recognition this is possible only when each is for the other what the other is for it, only when each in its own self through its own action, and again through the action of the other, achieves this pure abstraction of being-for-self.
>
> *(1977, p. 113)*

The struggle for recognition is the birth of the Ethical Life in which we come to know and act in accordance with a shared substance. We finally perceive ourselves as united in an identity despite our many differences and disagreements. Negation becomes a matter of political change in which our needs are met through our political institutions and private endeavours. Our practical reason is coalesced into a system of needs which contains all our individual ends embodied in the state.

Hegel (2003) thereby argues that Right (from the German "Recht" meaning moral justification and legal rights) is, therefore, free will and the system of right is actualized freedom. He refers to the institutional system built from sharing a common society and common laws. These laws train us to respect and uphold a conception of rational mutual relations for the good of ourselves and society. I recognize myself in my responsibilities to others, and I recognize others in their responsibilities to me and vice versa. If there is a conflict within these obligations, it marks an institutional issue. These issues are often rational conflicts within the proposed ethical life of an agent, such as a person who is legally bound not to steal but is also incapable of feeding themselves within the bounds of the law. Hegel perceives this conflict as a tragedy or a conflict of two rights, which, once negated, marks the institutional progression of the state. Furthermore, unlike thinkers such as Locke (2016) or Kant, rights for Hegel can be positive obligations towards one another since the apprehension and exercise of each other's goods are the unity of the ethical life. Hence, Hegel sees no opposition between morality and law, placing a combination of these two as the foundation of ethical life. The less conflicted and more harmonious the system of needs, the better the ethical life and ways we understand our shared substance.

To sum up what we have learned from the Hegelian–Aristotelian perspective, we need to engage in mutual recognition. This recognition ought to lead us to perceive a shared substance through a dialectical movement of opposites, finally uniting into a single identity. This unity is made possible through social roles made to fulfil our desired goods. These social roles are imperfect since the recognition process is mediated through the power of the practical reasoners attempting to achieve their ends. This process leads the stronger to have more significant influence over the identity than the weaker, leading to the power imbalance to be continued through metaphysical/ideological means. The imbalance is corrected over time as the practical reasoners eventually reject their roles as unfitting after becoming discontent. The less powerful feel discontent with their (in)ability to choose for themselves and so reject the identity in relation to their unhappiness. The more powerful become unable to satisfy themselves since, even when they achieve themselves, they are unrecognized and so unfulfilled. The solution is found through the constant mediation of individuals within society as they struggle to achieve their practical ends through theoretical means, leading to a progression of essences until they finally discover their shared substance. Thus, it may be said that loving an alien is likely to be a bit of a struggle but worth it in the end.

We now have vital insight into how we may begin to approach the problems of ethics with extraterrestrial beings. First, we must acknowledge that our good and their good are the same; what aligns our goods is not a biological category, such as a species, but rather our being practical reasoners and moral agents. We must then recognize that their goods are necessarily valuable. Without acknowledging all such goods, we will be as alien to ourselves as they are to us. We must attempt to build a relationship with them and seek these goods together. We cannot intuit inherently our own goods and must look to the other to negate our implicit mistakes while doing the same for them. Both parties will recognize each other in reference to this negation, marking the opportunity to build a shared identity. This identity must be structured and trained through political means so that we may continue to develop our consideration of ourselves and find a stable and complete substance. In order to reach these goals, we are presented with our first problem: Will we be able to perceive aliens as practical reasoners in the first place?

In multiple works of science fiction, writers have presented us with creatures and objects so foreign that we do not perceive them for what they are. Examples include silicone, artificial, extradimensional life forms and a myriad of others which live outside our usual perceptual range. However, beyond science fiction, we might have already overlooked alien life due to its differences and our limited modes of inquiry. One such example may be the Viking landers, which performed experiments on Martian soil in the 1970s. After experimentation, it was concluded that there was no sign of life. However, recent discoveries of extremophilic organisms have thrown this into doubt, with Houtkooper, J.M. and Schulze-Makuch, D. (2007) suggesting we may have even killed the life present with our analysis simply by exposing it to water. First contact with an alien species presents unique difficulties, and we must find a way to communicate our animality to one another despite the possibility of huge hurdles.

Regardless of *how* we communicate, the Aristotelian approach suggests that *what* we are to communicate is our needs; in other words, both parties need to demonstrate and perceive each other as practical reasoners. In order to do this, both parties must be willing to analyze the other to discover specific functions which indicate animality. Such functions must indicate the capacity of an internal experience exhibited through observation alone. These actions could be understood as a being attempting to achieve some form of end and exhibiting specific behaviour. For example, we might perceive that a being experiences pain or pleasure through observing them move away or towards a specific stimulus of their own volition. Such behaviour indicates a positive and negative experience, which Aristotle, Aquinas, and Hegel take to be the result of the flourishing or frustration of a being's nature.

It would be further bolstered if we observed that certain conditions harm the longevity and well-being of a creature or otherwise sustain the life of the creature in question. These patterns indicate that a being is consciously aware of its surroundings and altering its behaviour to achieve an externalized goal. The inference is that upon attempting the behaviour, the creature receives feedback in the form of an experienced positive or negative stimulus which it could then use to alter its actions, such as experiences of pain or pleasure. However, this conclusion is not agreed upon unilaterally; Wittgensteinian inspired thinkers such as Leahy (1994), who wrote *Against Liberation*, and Hartnack (1972), author of the essay "On Thinking", both disagree stating that language is a prerequisite for thinking, a prerequisite which most animals do not possess. However, this claim can be seen as early as Rene Descartes in the 17th century who also rejected animal's having thoughts altogether (Descartes, 2006). However, the thinkers do have differing reasons for their claims. Descartes denied animal consciousness on metaphysical grounds, arguing that animals are mere automatons entirely lacking subjective qualitative experiences. Leahy, on the other hand, doesn't deny consciousness outright. He instead takes animals to be bundles of sense data and behaviour, with no powers of recognition or conceptualization. Similarly, Hartnack argues that thought requires language, since a thought implies the power of *stating* some knowledge the thinker has. However, all these views share a common element: Consciousness results from reflection or at least the capacity to do so. It is also a commonality in some papers in neuroscience which attribute consciousness to higher-order mental abilities, typically citing a need to unite cognitive faculties under a self-reflective whole.

MacIntyre, who is a contemporary Aristotelian, answers these positions by demonstrating that our ability to produce language is first reliant upon our ability to have and hold prelinguistic beliefs:

> For human beings too there is an elementary prelinguistic distinction between truth and falsity embodied in those changes of belief that arise immediately from our perceptions of changes in the world and issue in changes in our activity... The acquisition of language enables us to characterise and to reflect upon our prelinguistic and nonlinguistic distinction-making in quite new ways, but there is an important continuity between the prelinguistic and linguistic capacities. The former provide matter for characterisation by exercise of the latter and in so doing place constraints upon the application of the concepts of truth and falsity that are provided by and in language.
>
> *(2002, pp. 36–37)*

In other words, our capacity to form propositions linguistically is reliant upon an already prelinguistic distinction made available through our observations and nonlinguistic beliefs. MacIntyre is thus defending the Aristotelian relation of the animal soul to the rational soul. He argues that the rational soul is only a modification of the animal soul, as the reflective reason is fundamentally the same practical reasoning displayed when an animal attempts to achieve an external goal, such as when they seek food or shelter. However, the difference is that this reason is reflected back upon itself such that its goal is that which is being contemplated, thereby making the object of analysis the animal itself. This means that the reflective reason is not an entirely new faculty, instead we should understand the rational soul as an evolution built upon the animal soul. MacIntyre even goes so far as to say that this view is most compatible with an evolutionary position, since the alternative seems to create a substantial distinction in kind between Humans and other non-human animals which is not supported by empirical evidence. Instead, he argues that the evidence suggests a spectrum of intelligence and linguistic skills, which we can see a pattern of increasing iterations, ranging from the more simplistic modes of communication between dogs through to the more complex examples of chimpanzees and dolphins onto Humans.

Thus, we should be able to identify alien modes of behaviour through pain responses and possibly even modes of pre- or nonlinguistic communication, at least in principle. However, every case will be subject to the specific nature and means of observation available; this is at least cause for some hope. Albeit celebration may very well be premature, since even if we perceive their animality, their needs, and vulnerability, we might disregard it as inconsequential to our moral decision making.

Even a cursory overview of history provides ample evidence that human beings have a tendency towards discrimination and prejudice aimed at minorities and those deemed "other". Often such discrimination has culminated in the denial of the basic dignity and moral equality of its victims, resulting in unspeakable atrocities. The Atlantic slave trade and the Nazi Holocaust are both examples of the evils human beings will commit against one another. At the time, these abuses were justified through moral reasoning which we have, thankfully, later determined to be inadequate. Can we say we would not be guilty of the same regarding extraterrestrials? Many of us already believe a cognitive dissonance exists between moral principles and our actions. One such example concerns the treatment of non-human animals. These beings we usually admit feel pain and pleasure; we often keep them as pets and form close relations with them, yet many of us choose to consume them and their products despite it often leading to their exploitation and death. Can we honestly say that we might not share more in common with these terrestrial creatures than an extraterrestrial? Such alien creatures may not share the same vital structures, behaviours, or experiences beyond a fundamental imperative towards flourishing and away from destruction.

What we are now touching on is ideology. The problems of ideology have been remarked upon at least as early as the French Enlightenment thinker Jean-Jacques Rousseau (1712–1778) (Rousseau, 2019), who believed that man was socially conditioned by his society in a way that was degrading to his nature. Hegel, who agreed with Rousseau, expounded upon this in the Lord and Bondsman relationship mentioned earlier, arguing that this was primarily due to an imbalance of power, ultimately resulting in a system of right being one sided. However, Hegel believed we would ultimately resolve all our ideological biases, ridding ourselves of these mistaken modes of judgement, albeit at the end of history. The question is, how ought we free ourselves from a psychological bias that corrupts our judgement? Worse still, how do two parties with separate forms of reason eventually resolve their competing forms of right?

Part 4: Political Prejudice and Complete Community

We could easily imagine a scenario in which we are presented with alien life which does not fit our criteria for moral consideration; likewise, we might also present in the same manner to said aliens. For example, we may both judge each other as inferior regarding our own political and moral systems. How will we then resolve this? Hegel argues that the resolution will not come through mere reflective reason alone but through the conflict within practical reason, which leads to tragedy. Tragedy for Hegel is a moment in which one cannot help but do the wrong thing, since every seemingly right action violates a set of rights or principles of either party. An example is a mother stealing to feed her child; she has a duty as a mother to care for her child and a duty as a citizen to refrain from stealing. What ought she do? Either situation presents itself with the mother having to act against her own ethical life to maintain herself; this means she is in a state of unfreedom, or in Aristotelian terms, she is prevented from flourishing. The fault Hegel explains is not with the mother but with the state. The state for Hegel is built upon what he calls a system of needs. This system for our purposes is the collective practical reasoning of the individuals who participate; in their attempts to achieve well-being and happiness, they construct the institutions which regulate their behaviour towards a harmonious relationship. In the case of extraterrestrials, the risk of conflicting interests, behaviours, and beliefs becomes even starker. That said, in all states, the institutions are imperfect and, eventually, tragedy will arise; the question is, what do we do about this seemingly inevitable tragedy? Can we learn anything from our own political turmoil to help us answer this crucial question?

MacIntyre argues that what we can perceive within our political systems today is a form of incommensurability, an inability to communicate, leading to greater conflict. MacIntyre uses the case study of the American political parties to demonstrate a variation in our forms of rationality, which means that when presented with an identical scenario, it leads us to form contrary moral judgments. What are we to do since our mode of inquiry leads us to judge moral and political reality vastly differently? MacIntyre proposes an unorthodox solution for philosophy; he argues there cannot be a resolution internal to any form of rationality without embracing an empathetic view towards an external critique (MacIntyre, 1988). This view should be adopted and understood as much as possible by the tradition's adherents through empathy in which the thinker embraces the thinking of the alien tradition as embodied in their life, history, and institutions relative to their attempts to achieve the good.

However, such a mutual understanding is not an easy goal; MacIntyre outlines two tasks the thinker must engage with: the internal critique, such as the ongoing debates within their tradition and the external critique of its rivals, marking a substantial distinction. The latter, he argues, is the most difficult and requires adopting a "second first language" in which an individual embraces an imaginative act of empathy placing themselves in the history and community of their rational rival. One such task, he notes, is not a true embodiment of being a native speaker but is a step towards recognizing what it means to be a native speaker within that community; he says:

> To possess the concepts of an alien culture in this secondary mode, informed by conceptual imagination, differs in important ways from possessing the concepts which are genuinely one's own. For insofar as one disagrees upon whether or not a particular concept has application...because one's own conceptual scheme precludes its having application, one will only be able to deploy it in the way in which an actor speaking his part may say things which he or she does not in his or her own person believe...but this does mean that we cannot understand what it is to be and to believe within another tradition by acts of

empathetic conceptual imagination in some times of case; the limits of possibility here are those set by the kinds of untranslatability which were catalogued earlier.

(MacIntyre, 1988, p. 395)

MacIntyre's empathetic approach draws on the motivation to produce language in the first place. Since language would always form from prelinguistic communication and requires, as its basis, an act of mutual recognition. What is being recognized, MacIntyre argues, is the practical reasoning of the subject. Each conscious subject in their behaviours marks a goal-oriented way of being which gives us the means to form a game-like structure in which we can make moves which the other can respond to. MacIntyre is, therefore, introducing a Wittgensteinian language game consideration into his Aristotelian approach, noting that each response indicates something about the other player. In forgetting our reliance on practical reason, we become dogmatic and alienate ourselves from the theoretical reason made available within the language of the game. In recognizing this basic prelinguistic structure, we open ourselves up to perceiving prelinguistic belief states which underpin our modes of thinking. The game in which we play with others is at this level in which we use our empathetic capacities to gain vital information. This means that prelinguistic beings and even nonlinguistic humans are inherently valuable to us whether they return our recognition or not. Since our recognition of prelinguistic beliefs marks the bedrock of our linguistic beliefs, and so all our modes of inquiry. This is why, at its core, rationality must be seen to be united with animality in ethics; we are dependent rational animals, and in accepting our dependency on one another, we may use our reason to obtain our animalistic ends.

In many ways, this leaves us with the unsettling yet optimistic conclusion that when confronted with alien life, we share a common good, a single substantial end which dictates our possible individual ends. Since we all share it, we are all moved to seek it, giving us the possibility of diplomacy and eventual unity. We will likely be fraught with difficulties in perceiving practical reason and might even find ourselves disregarding it. Such disregard is something we are doing with prelinguistic animals, and it's something we must be willing to address with regard to alternative and alien forms of life. However, so long as we keep attempting to perceive the other, we have the possibility of resolving our dissonance. Furthermore, in our struggle for recognition, we might engage with higher and more complex forms of life with its own ideology. Such life forms may think incompatibly to our institutions; we may fall into direct conflict and reject each other as incoherent due to our incommensurability. Such incommensurability is our opportunity for charity and love. This moment demands our empathy and imagination to the highest degree, and we must hope our alien interlocutor will do the same.

References

Aquinas, T. (1911) *Summa Theologica*. Translated by Fathers of the English Dominican Province. London: R. & T. Washbourne (Part 1).

Aquinas, T. (1914) *Summa Theologica*. Translated by Fathers of the English Dominican Province. London: R. & T. Washbourne (Part 2).

Aristotle (1999) *Nicomachean Ethics*. 2nd ed. Translated by T. Irwin. Cambridge: Hackett Publishing Company.

Aristotle (2016a) *De anima*. Translated by C. Shields. Oxford: Clarendon Press.

Aristotle (2016b) *Metaphysics*. Translated by C.D.C. Reeve. Cambridge: Hackett Publishing Company.

Descartes, R. (2006) *A Discourse on the Method*. Translated by I. Maclean. Oxford: Oxford University Press.

Hartnack, J. (1972) 'On Thinking', *Mind*, 81(324), p. 10. Available at: www.jstor.org/stable/2252717

Hegel, G.W.F. (1977) *The Phenomenology of Spirit*. Translated by A.V. Miller. Oxford: Oxford University Press.

Hegel, G.W.F. (1991) *The Encyclopedia Logic*. 3rd ed. Translated by T.F. Geraets, W.A. Suchting, and H.S. Harris. Cambridge: Hackett Publishing Company (Encyclopaedia of Philosophical Sciences, 1).

Hegel, G.W.F. (2003) *Elements of the Philosophy of Right*. Cambridge: Cambridge University Press.

Hegel, G.W.F. (2010) *Philosophy of Mind*. Edited by M.J. Inwood. Translated by W. Wallace and A.V. Miller. Oxford: Oxford University Press.

Heraclitus (1987) *Fragments*. Translated by T.M. Robinson. Canada: University of Toronto Press.

Hobbes, T. (1998) *Leviathan*. Edited by J.C.A. Gaskin. Oxford: Oxford University Press.

The Holy Bible, Mathew 22:37–40. St. Joseph New Catholic Bible (2019), United States: Catholic Book Publishing Company.

Houtkooper, J.M. and Schulze-Makuch, D. (2007) 'A Possible Biogenic Origin for Hydrogen Peroxide on Mars: The Viking Results Reinterpreted', *International Journal of Astrobiology*, 6(2), pp. 147–152. Available at: https://doi.org/10.1017/S1473550407003746

Kant, I. (2000) *Critique of Pure Reason*. Translated by P. Guyer and A.W. Wood. Cambridge: Cambridge University Press.

Kant, I. (2018) *Groundwork for the Metaphysics of Morals*. Translated by A.W. Wood. London: Yale University Press.

Leahy, M.P.T. (1994) *Against Liberation: Putting Animals in Perspective*. London: Routledge.

Lewis, D. (1993) 'Many, but Almost One', in J. Bacon, C. Keith, and R. Llyod (eds.), *Ontology, Causality and Mind*. Cambridge: Cambridge University Press, p. 23.

Locke, J. (2016) *Second Treatise of Government and a Letter Concerning Tolerance*. Edited by M. Goldie. Oxford: Oxford University Press.

MacIntyre, A. (1988) *Whose Justice? Which Rationality?* Notre Dame: University of Notre Dame Press.

MacIntyre, A. (2002) *Dependent Rational Animals*. Illinois: Open Court Publishing.

Plato (2013) *The Republic*. Edited by G.R.F. Ferrari. Translated by T. Griffith. Cambridge: Cambridge University Press.

Rousseau, J.J. (2019) *On The Social Contract*. 2nd ed. Translated by D.A. Cress. Cambridge: Hackett Publishing Company.

Watts, A. (1972) *Mind over Mind* [MP3]. (Comparative Philosophy). Available at: https://alanwatts.org/transcripts/mind-over-mind/

5

MUSIC OF THE SPHERES

Could Aliens Understand Music?

Gregory Stacey

Overture

If we were to encounter intelligent aliens, how should we communicate with them? One suggestion found in both fictional and real attempts at extraterrestrial communication is that we might make use of music. In Spielberg's *Close Encounters of the Third Kind* (1977), for example, humans and aliens communicate through a sophisticated exchange of notes. Equally, the Voyager I and Voyager II spacecraft each carry "Golden Records" containing various media – including a diverse range of traditional, instrumental music – compiled by Carl Sagan with the aim of informing aliens about human life. Popular culture also includes depictions of alien species exhibiting harmonious co-operation through musical performance, such as the scene in *Star Wars* where a diverse band of aliens plays at the Mos Eisley Cantina.

On one level, attempting to communicate with aliens using music seems intuitive. Music-making typically involves the production of a sophisticated, ordered series of sounds. By listening to human music, aliens would be able to understand something of our intellectual and sensory capacities. Yet any co-ordinated set of actions or data perceptible to aliens – say, the performance of a dance or transmission of a message in morse code – would equally serve to inform them that we are embodied, intelligent creatures with the capacity for sense perception. Why, then, does music intuitively seem an especially fitting medium for extraterrestrial communication?

One tentative suggestion might be that music has a *universal* quality. Unlike language, many philosophers agree that music cannot communicate propositional content, and some deny that it can represent anything non-musical (see Davies, 1994, pp. 1–121; Scruton, 1999, pp. 119–139). However, music is more commonly regarded as a means of conveying ineffable mental states to others, or of expressing and even provoking emotions which can be experienced by individuals from different cultures. Music therefore affords an important means of communicating non-propositional content to others, which may allow them to share our experiences to some degree. As a result, one might judge that it is a useful way to try to give aliens some sense of human experience. It might prove especially important

DOI: 10.4324/9781003440130-8

if we are unlikely to be able to translate any alien languages, as another contributor to this volume argues. Additionally, there is something inherently explorative about music, which echoes human efforts to explore physical space. Insofar as we can express our inner lives through music, its performance affords a chance to project the "virtual" space of human experience into physical space. In live performances of music, this projection nevertheless results from and draws attention to the location and bodily actions of musicians. As Ken McLeod (2003, p. 338) writes, "music takes us outside of our bodies and place while simultaneously reminding us of our location and what it means to live there". Given music's distinctive ability to facilitate cross-cultural communication and to display our internal lives in physical space, it seems fitting that music should form part of our efforts to contact extraterrestrial life.

In this chapter, I evaluate the intuition that music is an appropriate medium of communication with aliens. I focus on a fundamental question: Is it probable that aliens can *understand* human music? For if aliens cannot appreciate, one might assume that there is little point in using it to contact them. To avoid unnecessary complications, I will principally consider "pure" music – that is, music without lyrics, performed in a concert setting.[1] Before attempting to answer the question just posed, I should first clarify what "understanding music" involves. As an initial suggestion, we might say that this involves understanding music (i.e., certain collections of physical sounds) *as music*. Yet this might seem unhelpful without a clear definition of music, and some philosophers despair of attempts to arrive at the latter.[2] Accordingly, I suggest that we begin by regarding music in inclusive terms as a form of sonic artefact which is found across human cultures and which can be identified by family resemblance. There are many uncontentious examples of musical forms (e.g., symphonies, songs, and ragas) and types of performance (in concert, or as an accompaniment to dance, or during religious services), alongside putative examples which are more contentious (e.g., John Cage's 4'33, or the ringing of "methods" on church bells). On this rough conception, it is possible that aliens also possess music if they produce and listen to artefacts which we would readily describe as "music" given their resemblance to human music.

Musical understanding comes in different species and degrees. Here, I highlight three degrees of musical knowledge, without claiming to present an exhaustive taxonomy. First, there is the basic ability to distinguish music from other sounds. As Stephen Davies notes, this is capacity is multifaceted. It involves the ability (i) to differentiate musical sounds from sounds not produced by musicians, (ii) to separate sounds produced by musicians which are and are not part of the musical performance, and (iii) to distinguish overlapping musical pieces or performances (Davies, 2011, p. 89). Arguably, it also requires the imaginative ability to hear musical sounds as unity, so that both individual chords, sections of music, and pieces are heard as unities; I return to this point below. Second, there is the ability to appreciate music in musical terms, by which I mean the ability to identify those distinctive musical features in virtue of which different pieces and performances are individuated. Some of these features (like key, time-signature, and melody) are essential to the piece being performed, whilst others (such as tempo and timbre) are, at most, essential only to token performances. Philosophers debate whether this form understanding must be mediated by the technical terms provided by music theory in order for it to be conceptual and to reach a high degree of sophistication. But I here assume with Davies (2011, pp. 100–105) that listeners who are not formally educated in music theory can typically deploy folk-theoretic terms and concepts to reach nuanced, conceptual knowledge of music and its properties. Similarly, although aliens would initially lack concepts provided by human musical theory when first listening to human music, this need not necessarily prevent them from appreciating music's distinctive features through their own concepts, which might be translatable into human musical concepts. Finally, at perhaps the most abstract level, there is the ability to understand music as an artefact in its cultural setting. For instance, one might recognize a melody as an imitation of another or recognize a composer's achievement in inventing or expanding a genre. This form of understanding is also conceptual and often requires knowledge of the intentions of the composer and/or performer.

In this chapter, I will be most interested in exploring whether aliens are likely to be able to understand music in the first two senses mentioned above; that is, whether it is probable that they can identify music and have some appreciation of its distinctive musical features. If it is unlikely that any aliens which we might encounter have either of these abilities, there seems little point in attempting to communicate with them through music rather than other media.[3] Admittedly, it is vanishingly improbable that any individual alien species with whom we might make contact would have *precisely* the same language or concept of "music" as humans, but I take it that if aliens have a natural capacity for "perceptual" musical understanding and the broader ability to use and communicate concepts, they would, in principle, be capable of acquiring and using the human concept "music" and other associated concepts (e.g., "pitch", "rhythm", and "expression"). Accordingly, I will not dwell at length on the probability of aliens possessing the capacity to form, process, or learn human concepts.

So much for the nature of musical understanding. But what would it mean to ask whether it is *probable* that aliens understand music? Probability should here be understood as epistemic rather than statistical or metaphysical. Epistemic probability concerns the degree of credence with which one should rationally believe a proposition, given some or all of one's available evidence.[4] In this case, the probability which we are seeking is (roughly) the degree to which we should rationally believe that aliens can understand music, given our knowledge of the nature of music and the *a priori* probability of aliens possessing the various capacities required for musical understanding. More colloquially, this chapter investigates how surprised we should be if we discover aliens who can understand music. Second, I note that the relevant probability is the probability that any *particular* alien species which we encounter can understanding music, abstracting from any fine-grained prior knowledge which we might have of their cognitive capacities or cultural life, besides that which is specified below. Perhaps if there is a large enough universe or multiverse, it is highly probable *a priori* that there are aliens capable of understanding music. Maybe it is highly likely that there is a "twin earth", as envisaged by Hilary Putnam, whose humanoid denizens listen to alien equivalents of Holst or Bowie whilst sipping glasses of XYZ. However, if such aliens are very unusual, we should be very surprised that any token alien species which we encounter can appreciate music.

Nevertheless, I will assume that we are only interested in the probability that aliens who have some hope of appreciating music can do so in fact. If humans ever encounter extraterrestrial creatures which are evidently incapable of musical understanding, we will not attempt to communicate with them through music, nor are they the intended audience of Sagan's "Golden Record". To constitute a prospective audience for human music, aliens must have various abilities, most fundamentally, the capacity to "hear" human music, that is, the ability to somehow physically perceive sounds at the frequencies typical of human music. They must also have sufficient cognitive capacity to perceive the relationship between the sounds which make up music and to perceive musical properties in those sounds. Finally, some philosophers believe that musical understanding requires the ability to demonstrate one's understanding to others (Huovinen, 2008, pp. 315–320). I want to minimize assumptions in this in this chapter, so I will not be assuming that the aliens in question are those capable of showing us that they understand music. However, if music is used as a medium to communicate with aliens, it may not be useful unless aliens cannot eventually demonstrate their musical appreciation to us.

But how much should we expect any relevant alien species which we might encounter to differ from humans in other ways, as regards the capacity to appreciate human music? For the purposes of discussion, I will assume that aliens may have very different physiologies and brain structures from our own, because they may have evolved in extraterrestrial environments which bear little resemblance to earth. Equally, I will assume that alien "social" and "cultural" environments may be very different from our own. The precise respects in which the putative biological and cultural diversity of aliens may affect their capacity to arrive at musical understanding will become clear below.

However, one might suggest that alien biology and culture would necessarily be similar to terrestrial biology and culture. Evolutionary biologist Simon Conway Morris has suggested that the range of environments capable of supporting carbon-based life may be relatively limited (2011, pp. 557–561), which might reduce the biological variations in alien species. More broadly, Morris draws on his theory of "convergent evolution" to argue that intelligent carbon-based alien life would closely resemble human life (Morris, 2005, 2011, pp. 561–566). The key idea is that there are only a small number of optimal solutions to the problems of biological engineering which species inevitably face in the course of evolutionary history (e.g., how to perceive one's surroundings, move efficiently through a given environment, or reproduce). For this reason, various organs, such as eyes, have evolved independently in many species and environments. Notably, Morris even speculates that evolution might converge on the development of methods to transmit "complex information by vocalizations", such as song. If Morris is correct that alien physiology and social behaviour might be familiar to us, then perhaps my arguments below – aliens are unlikely to be able to appreciate human music – will be undermined. The reader should therefore evaluate their plausibility in light of their credence in Morris' suggestions. However, my analysis will retain considerable plausibility if it can be assumed that the physiology of intelligent aliens who we might encounter would likely differ from our own in at least the (relatively) modest ways in which the physiology of other intelligent terrestrial species differs from ours.

Many considerations bear on the overall epistemic probability of any proposition for a person or group. For reasons of space, in this chapter, I, therefore, simply advance two arguments that the epistemic probability that any aliens we might encounter have musical understanding is low, given extant philosophical accounts of the nature of music and our a priori judgements concerning the range of physical properties which aliens might exemplify. My first argument draws on recent philosophical analyses of music to suggest that aliens are unlikely to evince the first type of musical understanding introduced above: the ability to identify and distinguish music from other sounds. I contend that even if aliens can identify human music, music's nature nevertheless unlikely that they can understand music in the second sense given above, by appreciating its distinctive properties. My second argument further supports this conclusion by considering music's expression of emotion. I canvas several prominent accounts of the expressive nature of music, arguing that on each, aliens are likely incapable of understanding the expression of emotion in music. I conclude with a brief reflection on ways in which there might be some reason to communicate with aliens using music, even if prospective alien audiences probably lack the capacity for musical understanding.

Musical Understanding and Imagination

Many philosophers are reluctant to define music. But when they attempt such definitions, they sometimes refer to distinctively "musical" features of sound, such as pitch and rhythm. For example, Andrew Kania's definition specifies that, "Music is (1) sounds, (2) intentionally produced or organized (3) either (a) to have some basic musical feature, such as pitch or rhythm, or (b) to be listened to for such features" (2011, p. 28). It might seem unfortunately circular to define music in terms of "musical" features, but Kania's proposal highlights a difficulty inherent in providing a non-circular definition. Plausibly, there is a "level distinction" between music and sound. That is, whilst pieces of music are sonic artefacts which have specifiable sonic properties, such as volume and pitch, music cannot be differentiated from other types of sound by any distinguishing set of sonic properties. In other words, the presence of sound and sonic properties is necessary but insufficient for the existence and identification of music.

This distinction between sound and music is clearly elucidated by Roger Scruton. Scruton distinguishes three levels of properties and objects at play in music. Most obviously, there are the directional physical vibrations and associate changes in pressure which physicists might call sound. However, Scruton prefers to reserve the term "sound" for what he terms "secondary objects" – that is, intentional objects of sense perception which exist insofar as observers are disposed to perceive them to exist – generated by human attention to those vibrations. Nevertheless, one can be mistaken about the existence or properties of sound, because the latter are determined by common human perceptive capacities. As Scruton (2009b, p. 59) explains, the "existence, nature, and qualities [of second objects] are all determined by how things appear to the normal observer". Other examples of secondary objects include rainbows and smells. To support his analysis, Scruton argues that when non-scientists describe sounds, they refer to objects entirely different from sound waves (Scruton, 1999, pp. 1–9, 2009b, pp. 57–62). One reason for thinking this is the phenomenon of "acousmatic" experience, in which one listens to a sound and experiences it as a sound without paying any attention to its physical source or any of its physical properties, such when one hears a voice from behind a screen or a symphony played on the radio. The knowledge of sounds which is gained through acousmatic experience is plausibly knowledge by acquaintance: "a knowledge of 'what it is like', which is inseparable from the experience that delivers it" (Scruton, 1999, p. 1). It follows that deaf people – insofar as they have never had such acousmatic experiences – cannot hear sounds even if they can use instruments to sense the presence of physical vibrations and thereby detect the presence of sounds and acoustic properties such as pitch. As secondary objects, sounds are therefore wholly separable from their physical causes (Scruton, 2009a, p. 21).

One might think that Scruton would therefore describe the "musical" properties of sounds such as tone and pitch as secondary qualities, because they ostensibly belong to secondary objects. Yet Scruton describes these as "tertiary qualities". His motivation for devising another ontological category to describe properties such as tone is that they "are perceived only by rational beings, and only through a certain exercise of imagination, involving the transfer of concepts from another sphere" (Scruton, 1999, p. 94). As Scruton puts it elsewhere, the perception of musical properties is essentially metaphorical, insofar as it involves hearing sound with a "double intentionality", such that in hearing music, one perceives (secondary) properties which the sound really possesses together with properties which one realizes the sound does not actually possess but are applied to it "metaphorically" (Scruton, 2009a, pp. 43–45).

If this is hard to follow, we might consider why one might hold that descriptions of musical pitch (i.e., notes being "higher or lower") or descriptions of musical "movement" (which underlies much description of the progression of lines or harmonies) are somehow imaginative or non-literal. As Kania notes, there is no obvious physical correlation between pitch qua frequency of vibration and "height". A fortiori notes themselves do not "move" anywhere (Kania, 2015, p. 158). If we hear something "moving" in music, it cannot be the sounds (even, qua secondary objects) themselves, because a note's pitch is plausibly one of its essential properties. Scruton (1983, p. 114) therefore claims that "Understanding music involves the creation of an intentional world, in which inert sounds are transfigured into movements, harmonies, rhythms, – metaphorical gestures in a metaphorical space".

Not all philosophers accept this analysis. Malcolm Budd was once puzzled by the suggestion that a metaphor might form part of an experience rather than being applied to the latter on reflection (Budd, 1985, p. 242). In a later article, he worried that one cannot give an acceptable interpretation of Scruton's suggestion that when (for example) we hear a melody, we necessarily perceive sound as moving (Budd, 2003, pp. 215–220). For instance, as suggested above, it is implausible that we imagine individual tones moving through musical "space", because tones have their pitch as an essential property (ibid., p. 216). He therefore argues that such "metaphorical" descriptions of musical properties are strictly eliminable. Yet as Kania (2015, p. 162) notes, this approach struggles to account for some elements of pitch, such as octave equivalence. Others have proposed various ways to understand Scruton's suggestion that musical properties are "metaphorical", language about which Scruton himself became more cautious in his later years (see Kania, 2015, pp. 158–165; Scruton, 2009a, pp. 44–48).

But at its broadest, we might characterize Scruton's analysis as the persuasive suggestion that when we describe sounds as possessing musical properties, such as pitch and movement, we "imaginatively" attribute these properties to the sounds. These properties are neither possessed by any physical events or objects which cause sounds nor by the sounds themselves (i.e., individual secondary objects), which jointly constitute a piece of music. In this way, one might hold that merely hearing chords, lines of music, or even pieces themselves as unities of sound is itself a matter of imaginative perception. As Scruton writes, "Of course, we *hear* a chord as a single musical object: but that is the result of our musical understanding. It is not a feature of the spatial distribution of sounds" (1983, p. 98).

Clearly, the precise way in which music is the product of metaphorical or imaginative perception is contentious. Nevertheless, if musical understanding is *somehow* essentially imaginative, there is a simple argument that it is unlikely that aliens possess musical understanding in either the first or second senses given above (i.e., the ability to identify sound as music and to appreciate its musical properties). This is just that whilst our knowledge of the origins and causes of our imaginative capacities is limited, they likely depend (*pace*, mind–body dualism) on our brain structure. Even aliens who otherwise resemble us closely might – for all we know – very easily have entirely different evolutionary histories and brain structures (etc.), and so also very different imaginative capacities from us. Moreover, one's imaginative capacities and proclivities depend on one's cultural milieu. But aliens might also have very different cultures to our own. It therefore seems probable that (i) aliens would not imaginatively perceive the sounds which constitute human music as unities which are separate from non-musical sounds, and/or (ii) they would not perceive the sounds which comprise music as possessing those musical properties which humans typically perceive them to possess.

With respect to the latter point, it is noteworthy that although there is considerable overlap in the perception of pitch across human cultures (Davies, 1994, pp. 231–232), different cultures describe pitch in different "metaphorical" terms. Thus, whereas Westerners intuitively perceive pitch in terms of height, other cultures use the imagery of size or colour (ibid.). Moreover, some "metaphorical" musical concepts are more obviously dependent on cultural and environmental contexts, such as the descriptions of melodic intervals used by the Kaluli of Papua New Guinea, which are based on the properties of waterfalls (Feld, 1981). Accordingly, given the potentially very wide range of histories and brain structures which creatures may possess, it is epistemically probable that humans and any aliens we encounter would imaginatively perceive sound in different ways. But if aliens do not imaginatively perceive sound as we do, they probably lack a natural or intuitive capacity to understand music in the second sense specified above; that is, they cannot easily perceive music's distinctive features such as rhythm, pitch, and harmonic progression.

One might object that even if aliens are unlikely to imaginatively perceive the sounds which constitute human music in the ways in which we do, they may imaginatively perceive those sounds in other ways. If aliens perceive pieces of human music as unified wholes, we might grant that they have musical understanding of the latter in the first sense specified above, and perhaps also in the second sense. For instance, perhaps aliens could somehow imaginatively perceive pitch, but somehow (metaphorically) associate different pitches with different positions on a non-spatial spectrum of tastes or smells. Or more prosaically, they might perceive melodies as unities of sound extended through time, but not perceive any "movement" between notes. In these ways, aliens could both (i) identify sound as human music and (ii) have some understanding of its musical features without perceiving all the musical properties which humans perceive, or at least without perceiving those properties in the same way.

Yet if aliens were to appreciate music in this way, they would likely not share an understanding of a musical piece's essential properties with humans, and so they could not appreciate pieces of human music as distinctively human sonic artefacts. To see this, consider that if we think of music as sound with "tertiary properties", we can think of individual pieces or performances as "tertiary objects" (Scruton, 1999, p. 161). Some tertiary properties will be essential to those pieces or performances. However, tertiary objects and their properties are, like secondary objects, constituted by "how things appear to the normal observer". So, if aliens cannot perceive the same tertiary properties in sound as humans, then they cannot perceive the same tertiary objects as humans (i.e., pieces of human music) when they hear the music being played. As Scruton (ibid.) explains, "Only a being with certain intellectual and imaginative capacities can hear music, and these are precisely the capacities required for the perception of tertiary qualities". By analogy, we can imagine that if an alien language were fortuitously similar to the language in an English novel, an alien might be able to "read" the novel. Nevertheless, if the meaning of the text in the alien language was completely different from its meaning in English, we would not say that aliens and humans could read the same novel without translation, even if (improbably) the text's meaning in an alien language displayed sufficient unity for it to count as a novel.

Musical Understanding and Emotion

Another central topic in the philosophy of music concerns the relationship between music and emotion, since emotion is often considered central to our experience of music. I will now explore some theories which explain how music is expressed in music. Although I will not have time to examine every extant theory, I now argue that on three of the most prominent current theories, it is unlikely that any individual alien species which we might encounter will easily or naturally appreciate the expression of emotion in music. Members of such species will therefore lack one important facet of the second type of musical understanding specified above.

Listeners report emotional reactions to music and perceive emotion as somehow present or represented in the music itself. To capture the latter experience, philosophers often talk about music "expressing" emotion. However, following Eduard Hanslick, some "formalists", such as Edward Zangwill (2004), have denied that music can express emotion for several reasons. First, they allege that emotions necessarily have intentional objects. Thus, for instance, one cannot feel fear without being scared of something. Second, they argue that one can give an exhaustive description of any musical piece without using emotional language. Finally, they note the strangeness of attributing to emotion to music, which is not itself capable of feeling. Given the latter objection, we should draw a distinction which, as explained later, some philosophers use to make claims that music expresses emotion intelligible. Following Peter Kivy, we can distinguish "expressing" from "being expressive of" (1989, p. 12–17). An action or property "expresses" emotion if it indicates the presence of emotion in some subject. For instance, my bent frame and tearful face might express my grief. By contrast, if an action or property is "expressive of" an emotion, it represents an emotion which is not felt by any subject. For instance, a "weeping willow" might appear similarly mournful due to its "bowed" trunk, but to describe the willow as mournful is not to claim that it is capable of conscious experience.

There is no philosophical consensus regarding the nature of emotion. For instance, "cognitivists" and "non-cognitivists" have debated whether emotions necessarily involve judgements about intentional objects (e.g., fear "that ice is slippery", or joy "at reunion with friends"). However, a representative account of emotion is given by Jeremy Begbie (2011, p. 525), who identifies three components of emotion: conscious experience (e.g., "feeling" scared), bodily behaviour (cowering), and physiological activation (increased heartrate). For present purposes, we can adopt Begbie's analysis whilst noting that many philosophers recognize the existence of emotional states without intentional objects, which are commonly termed "moods".

Despite formalist animadversions, there is considerable experimental evidence that listeners can perceive emotion in music, even across cultures (Young, 2014, pp. 35–58). Accordingly, I assume here that music does express emotion and that appreciating music's capacity to express emotion is essential to musical understanding. This seems especially obvious in the case of pieces with lyrics, sung in contexts where the generation of emotion in listeners is important. How, for example, could one claim to understand Pergolisi's *Stabat Mater* without hearing grief in it or at least realizing that it in fact expresses grief? But the same holds true for even some pieces of "pure" music, such as Satie's melancholy *Gymnopédie No. 1* or the joyful fourth movement of Handel's *Music for the Royal Fireworks*. In such cases, it seems reasonable to say that the expression of particular emotions is an essential property of those pieces, if the latter are understood as "tertiary objects", which exist insofar as they are perceived by ordinary human listeners. However, I now argue that, according to many extant theories of music's capacity to express emotion, it is unlikely that aliens will be able to perceive emotion in music. To demonstrate this, I review three such theories in turn: the arousal theory, the contour theory, and the persona theory.

In their simple form, arousal theories argue that music expresses emotions or moods just by tending to provoke parallel emotional responses in listeners. Thus, Handel's *Music for the Royal Fireworks* expresses joy in virtue of its proclivity to evoke joyous excitement in an audience. They thereby offer implicit answers to the formalist claim that music cannot express emotion. Arousal theorists can argue that because emotions are located in listeners, rather than the music itself, they can easily have intentional objects. More generally, by locating emotion in audiences, rather than the music itself, arousal theories make the nature of the emotions in question less mysterious and can explain why music theory does not make reference to emotion. Whilst arousal accounts have drawn considerable criticism, *inter alia* because they locate emotion in listeners, rather than the music itself (see Scruton, 1999, pp. 144–146; Zangwill, 2004, pp. 33–40), they have also received recent elaborations (see Matravers, 2011). For example, Kendall Walton (1999) claims that music expresses emotion not only by arousing emotion but also by generating a feeling that there is someone or something else experiencing the emotion or mode which one is feeling. Understood in this or similar ways, the arousal theory appears closer to other accounts of musical expression, such as the "persona theory", which is detailed below.

Yet if music's arousal of emotion in listeners constitutes its expression of emotion (or aids perception of the latter), it is unlikely that aliens will easily be able to understand music's emotional expression. To see why, we should consider plausible accounts of how music can evoke the emotions which it expresses in listeners. Here, it helps to recall the embodied nature of emotions. Summarizing the results of contemporary psychology, Jenefer Robinson writes,

[M]usic directly induces bodily changes characteristic of particular moods, including autonomic changes, facial expressions, movements, gestures, and action tendencies. These bodily changes are experienced subjectively – and reported by listeners – as feelings of sadness, happiness, ebullience, anxiety, peacefulness, irritation, nervousness, etc. Crucially, the effect of the music is also to change energy and tension levels and in particular to produce the sorts of cognitive bias characteristic of particular moods.

(2009, p. 663)

If this analysis – for which Robinson (2009, pp. 664–670) provides empirical support– is broadly correct, then it seems unlikely that upon listening to music, aliens will tend to feel the *same* emotions as humans. This is because, as noted above, emotions are at least in part constituted by the forms of "bodily change", which Robinson describes here. But the tendency of one's body to respond to music in a particular way depends on the fine details of one's evolved biological features. Creatures with different bodily structures will therefore experience different physical – and consequently, emotional – reactions when listening to music.

This claim is not merely speculative; it is confirmed by recent work in animal psychology. In a recent paper, Charles T. Snowdon investigates the effects of music on animal well-being (Snowdon, 2021). He summarizes a range of experiments over the past four decades to assess the impact of human music on various species of animals, observing that according to current research, "music has a variety of effects that vary within and between species. There are few consistent results and it would be difficult to find any consistent effects of music, despite the large number of studies" (Snowdon, 2021, p. 6). In Snowdon's view, the reason for the inconsistent results is that researchers have often failed to consider the ways in which different types of human music will affect different species of animals given relevant biological differences. Snowdon notes that, first, animals often fail to hear human music – or hear it differently from humans – due to their auditory systems. More importantly, for present purposes, animal physiology also determines their physical reactions to music. One important factor which Snowdon highlights is heart rate. Music at about the tempo of an animal's heart rate appears to be calming, whereas comparable faster music is stressful or arousing. Snowdon was able to test his results on two species of animal: tamarins and domestic cats. Both species were largely indifferent to pieces which humans typically find calming or fear-inducing, respectively, but each species also responded with the appropriate emotions to pieces which were specially composed by David Teie to induce the relevant emotions.

Although, on Snowdon's telling, terrestrial animals react consistently to a range of musical features, including tempo, pitch, and dissonance, it is easy to imagine that their reactions might not be shared by alien species. For instance, whilst terrestrial animals have typically evolved to hear high-pitched and/or discordant sounds as indicative of danger or distress, aliens might hear such sounds in different ways. Equally, aliens might have a different heart rate to humans; indeed, they might possess several hearts, or none. We can scarcely envisage whether or how such creatures would naturally understand the expression of emotion through rhythm. In short, given that music generates emotional reactions in listeners by stimulating bodily changes, it is likely that aliens – if their bodies and environments are even modestly different from our own – would have very different emotional reactions to music, and therefore be unable to easily appreciate their expressive nature, if the arousal theory is true.

One might reply that if a simple arousal theory explains musical expression, then aliens could come to understand music's expression of emotion by attending to its emotional effects on human audiences. This seems possible in the abstract, but we should note some caveats. First, it would take a considerable degree of time and study for aliens to learn which emotions are elicited in human listeners by particular musical pieces. It would take aliens even longer to discover that, as Walton alleges, humans also imagine another person or persons experiencing the relevant emotion. Thus, aliens will not be able to appreciate the emotional nature of human music when they first encounter the latter. Second, it is one thing to notice that music is affecting another person or species, and another thing to be emotionally affected by music oneself. So, if arousal theories hold, aliens might have musical *understanding* (i.e., the ability to recognize music's expressive properties), but they would probably not have this understanding as mediated by (emotional) musical *experience*. Aliens would, therefore, have a different mode of musical understanding from humans.

Having seen that it is unlikely that aliens will be able to understand music's emotional nature of emotion on the arousal theory, we can now understand why that proposition is equally unlikely on more popular accounts, such as the "contour theory". According to the latter, as conceived by Peter Kivy (1989, pp. 46–83) and developed by Stephen Davies (1994, pp. 221–243), pieces are *expressive* of emotion because their perceptible features ("contours") resemble features or behaviours exhibited when humans experience emotional states. Several features of music have parallels in human emotional expression. First, the sound and speech of emotional people often resemble the volume and pitch of emotionally expressive music. Melancholy speech and music are both typically quiet and hesitant. By contrast, joyful speech and music are usually louder and often rise in pitch. Second, the speed of emotionally expressive speech and action also resembles the tempo of emotionally expressive music. For instance, sad people and sad music both move slowly. Finally, the patterns of tension and resolution in harmonic progressions often match the stress and relaxation which accompany and partly constitute changes in emotional states, as when anger resolves into peacefulness.

One advantage of contour theory is that it can locate emotion "in" music (rather than listeners), whilst avoiding some of the formalist objections that music cannot express emotion. Because contour theory holds that music is expressive *of* emotion, it does not make the mysterious claim that there is any real or imagined persona which experiences the emotions present in music. However, neither emotional nor musical "contours" possess intentional objects. Accordingly, on response to the formalist argument that emotions in music lack the intentional objects which are necessary for their individuation, some advocates of the contour theory hold that music can only be expressive of certain coursegrained moods rather than emotions proper.

Jerold Levinson's "persona theory" (Levinson, 2006) offers a refinement of contour theory. According to Levinson, rather than merely perceiving the emotional properties in music, listeners imaginatively perceive the music as the expression of occurrent emotions of a musical "persona". As Levinson himself puts it, "music expresses an emotion only to the extent that we are disposed to hear it as the expression of an emotion, although in a non-standard manner, by a person or personlike entity" (2006, p. 93).

The persona theory may seem implausible at first glance – listeners do not often claim to perceive a person "behind" emotionally expressive music. However, Levinson (2006, p. 104) clarifies that

> all that the theory requires is that they recognize the music as readily hearable as the expression of an emotion, even if, for one reason or another, they do not themselves give in to that inducement on a given occasion.

Moreover, Levinson's position has some advantages over the contour theory (Robinson, 2011, p. 105). On this view, we can simply say that music expresses emotions (of an imaginary persona), rather than holding that it is merely *expressive* of emotion. Further, on the persona theory, it is easy to see how pieces of music can express nuanced emotions with intentional objects – or, emotions which gradually develop throughout a piece. This simply requires a disposition in listeners to imaginatively perceive the musical persona as possessing such emotions.

Like the arousal theory, both the contour and persona theories have their critics (see the respective overviews provided by Robinson, 2011, pp. 202–204 and Levinson, 2006, pp. 103–108). But if either theory is true, it is unlikely that aliens we encounter will be capable of understanding music's emotional expression. According to both theories, we perceive emotion in music, in part, by noticing parallels between human properties or behaviours which express emotion and various properties of music. But just as it is *a priori* improbable that token pieces of music will generate the same emotional responses in humans and aliens, it is unlikely that emotions will have the same physical manifestations in humans and aliens. This is because the correlation between emotion and behaviour in organisms will depend on their environment, evolved physiology, and social norms. In all these respects, aliens might differ considerably from humans, meaning that behaviours, which particularly express emotions in humans and aliens, may likewise diverge. Perhaps, for instance, an alien species might have evolved to express sorrow through cries which rise in pitch or by moving rapidly. But if the contour or persona theories are correct, aliens will likely perceive music as expressing the emotions if they perceive a correspondence between its properties and the emotionally expressive properties or behaviours of their own species. Accordingly, aliens who express sadness through rapid movements will likely perceive pieces with a high tempo as expressive of sadness. They might, therefore, naturally hear Handel's *Music for the Royal Fireworks* as a dirge – and by contrast, Satie's *Gymnopédie No. 1* as a playful dance. Indeed, given the wide range of behaviours which might express emotion, it is somewhat unlikely that humans and the members of any individual alien species will perceive a token piece of music as expressive of the same emotions. The persona theory further specifies an important role for the *imagination*: for someone (Levinson, 2006, p. 95). But as noted earlier, since the imaginative capacities of aliens will also depend (*inter alia*) on alien physiology and culture, it is far from certain that any individual alien species would be disposed to imagine a *persona* which expresses its mental states through music.

As with my previous argument that aliens are likely to lack the imaginative capacities necessary to perceive sounds as music, my point here is not that if the theories of emotion in music I have explored are correct, then extraterrestrial audiences would be unlikely to perceive *any* emotion in human music. Notably, if their physiology and culture differ from our own, aliens may naturally experience human music as expressing different emotions from those which are perceived by human listeners. Yet as indicated above, some pieces of music are *essentially* expressive of particular emotions. If musical pieces are tertiary objects, then their essential properties are those which ordinary human listeners would typically hear when the relevant sounds are played. Accordingly, if aliens cannot naturally perceive the same emotions in a piece as human listeners, they may be unable to easily identify some of that piece's essential properties.

Perhaps aliens can come to understand the expressive properties of human music through the study of human emotions, emotionally expressive behaviour, and/or reactions to music. However, this will only be possible once they have engaged in extensive communication with humans. On first contact with aliens, they will probably not correctly identify the emotional properties of human music, even though we often assume that listeners from different human cultures are naturally able to perceive the latter.

Coda

We can now review this chapter's rather modest achievements. I have contended that it is unlikely that aliens understand human music. Yet I have only advanced two argument to that effect, and many other considerations may bear on the probability of that hypothesis. I have also made a substantial assumption throughout: that aliens can possess a wide range of physiologies and likewise have evolved in environments which are very different from our own. If one could show that any aliens who possess the auditory and cognitive capabilities necessary for musical understanding are likely to have similar bodies, environments, and cultures to our own, then one could go some way towards demonstrating that they would be able to understand human music.

Nonetheless, we can draw a provisional conclusion from my analysis. If the ability to appreciate music in either the first or second senses, outlined earlier, depends on the possession of a highly specific bodily form – perhaps in addition to a particular physical and cultural environment – it seems unlikely that even aliens who can hear and reflect on human music will be able to understand it. One might, therefore, judge that attempts to communicate with aliens musically are misguided insofar as they assume that aliens would naturally be able to comprehend human compositions.

But such a verdict would be hasty. Although I have suggested that alien audiences would be unlikely to be able to identify music sounds and properties, I have not claimed that they would be unlikely to perceive unified tertiary objects in the sounds which compose human music. Equally, I have not claimed that they would be unlikely to perceive those tertiary objects as emotionally expressive. In other words, if aliens have the capacity to listen to performances of human music and the intellectual ability necessary to form musical concepts, it seems plausible that they also possess the ability to hear "music" – that is, sonic tertiary objects. But the music (*qua* tertiary object) which aliens hear during a performance would likely differ from that heard by human listeners.

If this is correct, then humans and aliens might eventually have illuminating aesthetic discussions about performances of human music. When humans perceive a piece as composed of unified chords which "move" through musical space, aliens may experience the same sounds in completely different ways. Likewise, alien audiences might experience surprising emotional responses to human music. If aliens and humans can convey their respective experiences to one another – say, through the explanation of the metaphorical language which humans often use to describe musical experience – then we might learn new ways to imaginatively appreciate sound, or at least come to realize the contingency of our own musical experience. Rather than simply presenting aliens with human music and expecting them to appreciate its aesthetic qualities, we should perhaps be attentive to alien "music criticism". In other words, whilst we should not expect aliens to appreciate human music when we first make their acquaintance, dialogue with aliens about the aesthetics of sound might encourage us to re-evaluate how we understand music ourselves.

Notes

1 Admittedly, in their initial contact with humans, aliens are unlikely to hear music in a concert setting, and thus their first experience of a human piece will not strictly be an experience of "pure music". I bracket this point for present purposes, noting that upon first contact with humans, aliens might encounter human music without lyrics which was *composed* for performance in a concert setting (say, by recovering a "Golden Record").

2 On the difficulties of defining music, see Davies (2012) and McKeown-Green (2014).

3 I set aside the third sense of understanding (i.e., understanding music as a cultural artefact) on the assumption that for aliens to appreciate music in this sense, they must already have acquired extensive knowledge of human culture.

4 For a helpful overview of some competing understandings of epistemic probability, see Plantinga (1993, 137–158).

References

Begbie, J. (2011) 'Faithful Feelings: Music and Emotion in Worship', in *Resonant Witness: Conversations between Music and Theology*, edited by J. Begbie and S. Guthrie, pp. 323–354. Grand Rapids: Eerdmans.

Budd, M. (1985) 'Understanding Music', *Proceedings of the Aristotelian Society*, Supplementary Volume 59, pp. 233–248.

Budd, M. (2003) 'Musical Movement and Aesthetic Metaphors', *British Journal of Aesthetics* 43(3), pp. 209–223.

Davies, S. (1994) *Musical Meaning and Expression*, Ithaca, NY: Cornell University Press.

Davies, S. (2011) *Musical Understandings*, Oxford: Oxford University Press.

Davies, S. (2012) 'On Defining Music', *The Monist* 95(4), pp. 535–555.

Feld, S. (1981) 'Flow like a Waterfall: The Metaphors of Kaluli Music Theory', *Yearbook for Traditional Music* 13, pp. 22–47.

Huovinen, E. (2008) 'Levels and Kinds of Listeners' Musical Understanding', *British Journal of Aesthetics* 48(3), pp. 315–337.

Kania, A. (2011) 'Definition', in *The Routledge Companion to Philosophy and Music*, edited by T. Gracyk and A. Kania, pp. 1–13. New York: Routledge.

Kania, A. (2015) 'An Imaginative Theory of Musical Space and Movement', *British Journal of Aesthetics* 45(2), pp. 157–172.

Kivy, P. (1989) *Sound Sentiment: An Essay on the Musical Emotions*, Philadelphia: Temple University Press.

Levinson, J. (2006) 'Musical Expressiveness as Hearability-as-Expression', in *Contemplating Art: Essays in Aesthetics*, edited by J. Levinson, pp. 91–108. Oxford: Oxford University Press.

Matravers, D. (2011) 'Arousal Theories', in *The Routledge Companion to Philosophy and Music*, edited by T. Gracyk and A. Kania, pp. 212–223. New York: Routledge.

McKeown-Green, J. (2014) 'What Is Music? Is There a Definitive Answer?', *The Journal of Aesthetics and Art Criticism* 72(4), pp. 393–403.

McLeod, K. (2003) 'Space Oddities: Aliens, Futurism and Meaning in Popular Music', *Popular Music* 22(3), pp. 337–255.

Morris, S.C. (2005) 'Aliens like Us?', *Astronomy and Geophysics: The Journal of the Royal Astronomical Society* 46(4), pp. 4.24–4.26.

Morris, S.C. (2011) 'Predicting What Extra-Terrestrials Will Be Like: And Preparing for the Worst', *Philosophical Transactions of the Royal Society of London. Series A: Mathematical, Physical, and Engineering Sciences* 369, pp. 555–571.

Plantinga, A. (1993) *Warrant and Proper Function*, New York: Oxford University Press.

Robinson, J. (2009) 'Emotional Responses to Music: What Are They? How Do They Work? And Are They Relevant to Aesthetic Appreciation?', in *The Oxford Handbook of Philosophy of Emotion*, edited by P. Goldie, pp. 651–680. Oxford: Oxford University Press.

Robinson, J. (2011) 'Expression Theories', in *The Routledge Companion to Philosophy and Music*, edited by T. Gracyk and A. Kania, pp. 201–211. New York: Routledge.

Scruton, R. (1983) *The Aesthetic Understanding: Essays in the Philosophy of Art and Culture*, Manchester: Carcanet.

Scruton, R. (1999) *The Aesthetics of Music*, Oxford: Oxford University Press.

Scruton, R. (2009a) *Understanding Music: Philosophy and Interpretation*, London: Continuum.

Scruton, R. (2009b) 'Sounds as Secondary Objects and Pure Events', in *Sounds and Perception: New Philosophical Essays*, edited by M. Nudds and C. O'Callaghan, pp. 50–68. Oxford: Oxford University Press.

Snowdon, C. (2021) 'Animal Signals, Music and Emotional Well-Being', *Animals* 11(9), p. 2670.

Walton, K. (1999) 'Projectivism, Empathy, and Musical Tension', *Philosophical Topics* 26, pp. 407–440.

Young, J. (2014) *A Critique of Pure Music*, Oxford: Oxford University Press.

Zangwill, N. (2004) 'Against Emotion: Hanslick Was Right about Music', *British Journal of Aesthetics*, 44(1), pp. 29–43.

PART III

Communication

6

ALIEN WORDS

Some Difficulties with Pre-Dysonian SETI

Robert CB Miller

Introduction

SETI is the *search for extraterrestrial intelligence* (SETI). Modern SETI research began in 1959, with the publication in *Nature* of an article by Giuseppe Cocconi and Philip Morrison, 'Searching for Interstellar Communication' (Cocconi and Morrison, 1959), which argued for the use of radio telescopes to search for signals from alien civilizations. The result was a still continuing search and famously the receipt of the 'Wow! Signal' by *Ohio State University's Big Ear* radio telescope in 1977. This is believed to be a possible candidate for a genuine alien signal. *The SETI Institute* was established in 1984 and has been instrumental in SETI research ever since. The SETI programme, once launched, gained momentum and was supported for a time by the US Government. But this ceased in the early 1990s after controversy in Congress. Subsequently, it was supported by Paul Allen (co-founder of *Microsoft*) and the building of the *Allen Array* radio telescope for SETI. One interesting development was the launch of the *Breakthrough Listen* project, with $100m supported by the businessman Yuri Milner and Steven Hawking. Apart from the Wow! Signal, there has been no credible evidence of the receipt of an alien message.

This chapter will focus on attempts to detect and understand alien radio messages – pre-Dysonian SETI. Dysonian SETI was the term invented to describe the search for intelligent (i.e. language using) life as part of the search for extraterrestrial life. The term originates from Freeman Dyson's idea that 'advanced' alien civilizations might surround their suns with screens that would capture for their use much of the energy released. There was some excitement when it was thought for a time that 'Tabby's Star' (KIC 8462852), which is 1,470 light years from earth, might be one such. Thus, post-Dysonian SETI involves primarily the search from alien life from the detection of bio-signatures (and possibly techno-signatures). Pre-Dysonian SETI is restricted to the detection of radio (and other forms) of message.

The search for alien messages may well come to include laser, light, and transmissions across multiple parts of the electromagnetic spectrum. It is claimed that there are likely numerous civilizations in the Milky Way galaxy, some of which may be communicating

DOI: 10.4324/9781003440130-10

with us, with or without intending to. The types of messages which could be received from aliens fall into three categories: (1) Messages intended for us. These might be in the form of a signal from a 'beacon'. (2) Messages intended for other civilizations. (3) Messages that are communications between members of the same alien civilization – by-products. All of these present equal difficulties in translation.

SETI enthusiasts are optimistic about prospects of making 'Contact', and they accept some implausible theories to explain why contact has not yet been made. It needs to be explained why this is thought to be so important. It is claimed portentously that the discovery of alien intelligent life would be one of the greatest events in human history, and it would answer one of the most important questions facing human beings. For example, Dennis Overbye commented in the *New York Times* in an article reviewing SETI:

> A simple 'howdy,' a squeal or squawk, or an incomprehensible stream of numbers captured by one of the antennas here at the *University of California's Hat Creek Radio Observatory* would be enough to end our cosmic loneliness and change history, not to mention science. It would answer one of the most profound questions humans ask: Are we alone in the universe?
>
> *(2012)*

Without religious belief, many scientists (and for that matter other people too) claim to feel lonely in what appears to be a vast, disinterested, if not unfriendly, universe. Thus, SETI research is a means of finding meaning in the cosmos. The discovery of alien intelligent life will, it is thought, confirm that the universe is not the lonely empty place it appears to be. It is asserted that the discovery of alien intelligent life would explain better man's place in the universe. The following is an example of this assumption.

> Life is common in the universe. Emergence of intelligence and technology is, if not necessary, then at least a typical outcome of biological evolution throughout the Milky Way. A sizeable fraction of technologically advanced species seek to communicate with other intelligent creatures. It makes sense to listen for intentional radio optical messages from the depths of space and to transmit messages in return. It makes no sense to travel across interstellar distances or to expect such interstellar visitors. The best that we can hope for is the slow and benign exchange of messages – the greatest beneficiaries in such exchanges being the youngest new comers to the 'Galactic Club', including us humans.
>
> *(Quoted in Bradbury, Cirkovic, and Dvorsky, 2011, p. 156)*

Another motive for SETI is to find evidence to reinforce the Copernican doctrine that the universe is not centred upon earth and human kind. Thus, Jill Tarter former boss of the *SETI Institute* explained on TED:

> So, what would change everything? Well, this is the question the Edge Foundation asked this year, and four of the respondents said, 'SETI'. Why? Well to quote: 'The discovery of intelligent life beyond Earth would eradicate the loneliness and solipsism that has plagued our species since its inception. And it wouldn't simply change everything, it would change everything all at once.'[1]

But would the discovery of ETI make really much difference? No doubt the believers in the new scientific religion would see it as confirmation of their beliefs. But for others, it might be of no more significance than the discovery of a previously unknown exotic species on Earth.

SETI has had no success in more than 40 years of searching. The only possible candidate was the famous 1977 'Wow signal'. But this has never been repeated and it may be that an earthly source will be discovered. There have been no other plausible candidates and this has led to attempts to explain the failure to find a signal. These amount to variants of explanations of the Fermi Paradox.[2] These range from the possible to the grossly implausible – or just puerile. Still, so far, the search may not have been extensive enough for success. Of course, another possibility is that they do not exist and that we are alone in the Milky Way or even in the universe (Webb, 2015). Given the fact that scientists seem unable to account for the origin of life on Earth, this seems a strong possibility. One distinguished biochemist has claimed that scientists are 'clueless' about the origin of life (Tour, 2020), and the biologist Eugene Koomin has claimed that life on Earth is only possible because we live in an infinite multiverse (2012, p. 391). Consequently, it seems premature (if not a little reckless) to assert confidently that we cannot be alone.

Assumptions of Speech and Rationality

It is an unstated assumption that in discussing SETI by 'intelligence', it is meant language using. Language is a necessary condition for intelligence in the context of SETI. In other words, aliens must have language before they can send messages (or any other 'content' like music) to us or to anyone else. Without detecting a message, as we saw we might be able to detect intelligent aliens by the discovery of a 'bio-signature' or even a 'techno-signature' on an exo-planet – atmospheric pollution, for example. This assumes reasonably enough that technology is possible for creatures whose communications (if any) we have not detected.

In the main, human messages are related ultimately to human speech. All texts (including computer code) are representations of language. But we have no means of knowing that any alien message represents speech. Indeed, we have no means of knowing what (if anything) it represents. Without the limitations of human anatomy – alien speech (?) and language might include some thousands of different phonemes (or equivalent). But they might be the vehicle of alien music (or abstract art) rather than 'speech'. And language use suggests rationality. For us this implies such concepts as truth, falsity, consistency, logic, argument, freedom, and responsibility. In C. S. Lewis's science fiction novel, *Out of the Silent Planet*, the aliens are rational creatures that have freedom and responsibility. Still the forms of life of the *Hrossa*, *Seroni*, and *Pfifitriggi* are very different from each other and from humans (Lewis, 1938).

Human and Alien 'Civilizations'

One could argue that the 'loneliness' of SETI enthusiasts leads them to anthropomorphism. The universe (or at least the Milky Way) is imagined to contain numerous actual or incipient 'civilizations', many of which will suffer from the same angst as SETI enthusiasts. Professor Paul Davies supposes that an alien civilization might establish a 'beacon' (1995, p. 26). But again, this is to assume that aliens are like us and have the same anxieties and fears that we do.

Further the concept of 'civilization' is profoundly human. Civilization involves such concepts as language, contract, law, art, morality, religion, war, science, mathematics, architecture, engineering to give just a few examples. We have no reason to believe that any other life form will have replicated the extraordinary series of (providential (?)) 'changes and chances' that have brought human life and civilization to its current state. It is easy, like Stephen J. Gould in evolutionary theory, to imagine many equally possible but very different evolutionary trajectories and outcomes.

An unstated but important assumption of SETI is that humans and the aliens we might communicate with are fundamentally similar. But the different evolutionary paths of intelligent aliens might make them quite unlike humans in several respects, and it is unlikely that we share many of the same physical and intellectual powers, abilities, and motives. Thus, John Maynard Smith and Eors Szathmary have argued that over the 4bn years since the appearance of life on Earth, there have been eight major evolutionary transitions. These are the appearance of: (1) proto-cells; (2) chromosomes; (3) DNA genes and protein enzymes; (4) eukaryotic cells with nuclei and organelles; (5) sexual reproduction; (6) animals, plants, and fungi; (7) animal colonies, ants, wasps, etc.; and (8) human societies (Maynard Smith and Szathmary, 2000, p. 17).

Much of pre- and post-Dysonian SETI rests on the assumption that evolution is a progressive advance from stage to stage. But 'progress' is a profoundly human concept – and a recent one at that. It is assumed that civilizations 'progress', and knowledge of science and technology steadily improves without identifiable limit. This latter assumption seems based on very recent human experience. For we have no reason to believe that the last 500 years of scientific progress will continue on earth, or that it will be replicated elsewhere. One of the common assumptions of SETI theory is a belief in progress, both scientific and otherwise. It is always assumed that civilizations, ours included, have a course of technological development which is bound to be repeated by other 'civilizations'. The fact that human civilization has shown extraordinary development in terms of scientific knowledge, population growth, and wealth tells us little about what it will do in the future or what form it will take. The fact that 'progress' has characterized the last 500 years does not mean that an irresistible trend has been established for an indefinite period, or even for the immediate future. The growth of scientific knowledge has been extraordinary, but its sheer familiarity blinds us to just how amazing it is. Wittgenstein commented:

> What a curious attitude scientists have: 'We still don't know that; but it is knowable and it is only a matter of time before we get to know it!' As if that went without saying.
>
> *(1980, p. 40e)*

While for now we *may* expect scientific progress to continue. There is no telling how, if, or when, it will slow down, stop or even go into reverse. It is an unwarranted assumption to believe that science (or any alien equivalent) can continue far beyond what has been achieved so far. In any case, scientific progress is likely to be restricted by human intellectual powers, which are limited. Alien powers may be very different from ours and limited in different ways.

Thus, SETI depends on the assumption that aliens will be sufficiently like us for it to be possible for us to understand them and any messages they may send us. But it is worth considering how difficult it can be to understand human societies different from our own. Peter Winch in *Understanding a Primitive Society* argued that it was hard for us to understand a primitive society (Winch, 1964). We can only do so by analogy. He gave the example of an African tribe who believed their souls were contained in the small boxes that they carried. We could, he claimed, understand this by comparing it to the western practice of ascribing great value to wedding rings. How much more difficulty it would be to understand alien forms of life, conventions, and practices?

As we saw, there are puzzles and difficulties in assuming that alien life will have followed a similar evolutionary path to ours. Given the unique 4bn year evolutionary trajectory that resulted in human beings, it is likely that the evolution of aliens could have intellectual powers substantially different from humans. For example, intelligent aliens might be:

- ...completely unphased by the paradoxes of quantum mechanics.
- ...able to visualise a tesseract (a 4D cube).
- ...able to see that a group of objects totals 500 (or 5,000) items without counting.
- ...without the concept of a person.
- ...able to see well into the infra-red and ultra-violet. Maybe they could 'see' the whole electromagnetic spectrum.
- ...without contrasting subjective and objective experience.
- ...with a completely different experience of the passage of time. Their 'lives' may be lived far faster or slower than ours. Maybe they live in the eternal present.

These differences in alien intellectual powers would be reflected in any language they used and any maths and science they created. What is mysterious to us may be completely transparent to aliens and *vice versa*.

Translating an Alien Message

The most serious difficulties with the SETI project relate to translating and understanding any alien message. Suppose we were to receive a message emitted by an alien 'civilization'. Could we understand what it meant? *Prima facie*, this is a problem which could be quite easily solved given time and effort. Surely it could be argued that cryptanalysts should be able to decipher a message which the senders intended us to understand. If German Enigma messages could be deciphered when the Germans were doing their best to make them incomprehensible to us, then *en clair* alien messages should be easy. And this would be even truer of alien messages which the senders designed to be comprehensible to us.

Unfortunately, the translation of any alien message would not be like trying to decode an encrypted message sent from an Enigma machine. All messages which humans have so far received are from other human beings who use human language. Decoding ciphers is only possible because we share numerous concepts, interests and purposes with other human beings. In the Enigma example, the Bletchley Park code breakers knew that the Germans would be transmitting in encoded German and what sorts of military information to expect.

But in the case of an alien message, we would be most unlikely to share any concepts and assumptions with them and consequently we could not even begin to understand any message received. The need to share concepts before we can understand other people has been explained in different ways. Wittgenstein argued that language was only possible because human beings share 'forms of life' and linguistic conventions. For example, when somebody points it is assumed that one looks down the finger from the wrist rather than up the finger towards the forearm. Language has multiple uses and these are expressed in 'language games' which have different criteria for success. The later Wittgenstein was reacting against his former picture theory – the idea that language has meaning because it has the same structure as what it is describing, just as a picture has the same structure as that which it represents. It follows that since we do not share (or cannot know that we share) a form of life with aliens, then we cannot understand their language. To take the pointing example again, if aliens have no fingers or appendages, pointing is not a language game that they can play. Wittgenstein's famous remark, "If a lion could talk, we could not understand him" (1953/1968, p. 223e), illustrates the fact that since we do not share forms of life with lions, the ability to communicate with them is limited. What is true of lions will be all the more true of aliens.

Because we understand 'pointing' and all the other conventions that form part of language, we can point to things and activities to learn their names and roles in a foreign language. For example, Jane understands that thumping your chest and uttering a sound means that you are expressing your name. Only thus can she understand "Me Tarzan, you Jane." But there is no reason to suppose that an alien would understand the thumping and utterance convention – or indeed any other human convention. What is more Jane and Tarzan meet face to face, but that is most unlikely to be so in case of deciphering an alien message, which (if it is ever received) will be in some modulated form, electromagnetic or otherwise.

It might be argued that it would be no more difficult to learn an alien language than it would be to learn any other foreign language. But this is to ignore the fact that the ability to learn human languages is part, if Chomskyan linguistics is correct, of our biology and that the ability to acquire language is coded for in the human genome. Modern linguistics has discerned that all humans have the characteristic of 'discrete infinity' – the use of a finite number of elements to form an unlimited (infinite) number of sentences. This is achieved by a recursive set of rules, *Universal Grammar* (UG), which are built into human nature (Baker, 2001; Pinker, 1994). People with a defective FoxP2 gene have grammatical difficulties. An intriguing experiment has shown that humans can learn an artificial non-UG language but with difficulty (Musso, M., et al. 2003). It has been claimed that a very few 'defective' languages are not recursive (Everett, 2005, 2013, 2016, 2017). But these are not as expressive as ordinary recursive languages and *are* recursive at the sentential level. Alien language might not be like this. If (and this is a big if) recursive, aliens would most likely have their own version of UG. Another possibility is that they might achieve discrete infinity at the sentential level or by the use of another facet of their life form. Further, acquisition of another human language is always done in the context of shared lives, conventions, and understandings, as well as the actual (or virtual) presence of people. This suggests yet another obstacle in the way of understanding an alien message.

Translation: Ordinary and Extreme

Translation is usually from one known language to another. The person doing the translation must know both languages. In the prime instance, it is interpretation, for example, between French and English speech. The process can be illustrated as follows:

French <> English

Translation also involves texts – the ability to convert a French text into an English text. It is possible but unlikely that one could learn to translate a text in a foreign language without being able to speak the foreign language, or even their native language in the case of a deaf and mute person. Human languages can be expressed in three forms: speech, sign, and touch.[3] We have no means of knowing which of these (or other) methods aliens might use as their prime means of expressing their language.

In translating French speech into written English text, the process is as follows:

French Speech / Written French <> English Speech / Written English

De-cyphering

Cyphering and de-cyphering are done in texts. It involves languages known to the party attempting the de-cyphering. It is achieved by not disclosing publicly the rules for encoding and de-coding the encrypted text into and out of the base language. The task of the de-cryptographer is to discover (if he can) these rules and then to apply them.

Written / Spoken German > Encrypted German >DECRYPTION > Written German > Written / Spoken English

An unbreakable code is easy to construct but is very unwieldy to use, hence, they are not used often and usually codes which are very difficult to break but more convenient are used instead. Imagine, though, an unbreakable code set up between two parties who wish to keep their messages secret. They can do this by obtaining two identical copies of the same long rare book – not the Bible or Shakespeare. To send messages, the writer looks up the initial letter of his message at random in his edition of the book and crosses it out. He then identifies the letter in his encrypted message by giving the numbers of (a) the page, (b) the line, and (c) the number of the letter in the line. Thus, if the letter 'e' is the first letter of my message and it occurs on the seventh page in the eighth line and it is the ninth letter in the line, then the encoding of the 'e' will be 789. The protocol requires the deletion of the letter in the sender's edition of the book which indicates that it can never be used again. This means that every use of the letter 'e' in a message will have a unique encoded three-digit number.

Such a code will be unbreakable by any third party unless he has another copy of the same long rare book and knows its significance for deciphering the code. No amount of computing power effort of whatever kind can break the code. The significance of this for translating an alien message is that we would be in the position of a third party trying to break an unbreakable code. The reason is that we could have no possible access to what the message referred to, or what its elements signified.

Translating the Texts of Formerly Unknown Languages

It might be thought that as it has been possible to translate the texts of lost human languages, such as Egyptian hieroglyphs and Linear B, we should have, at least, a good chance of translating an alien language. But it is telling that attempts to decipher the writing of lost human languages have not always been successful and that success, when it has occurred, has been time consuming and laborious. Take four well-known examples where much effort and energy has led to the deciphering of previously unknown writing: ancient Egyptian, cuneiform, Mayan glyphs, and Linear B.

Translation of these scripts was impossible until the unknown language they represented was discovered or identified. (This is a necessary but not a sufficient condition – see the discussion of Rongorongo below.) This process of discovery or identification can be sometimes eased by the discovery of a 'Rosetta stone'. Thus, the Rosetta stone had three versions of the same decree: hieroglyphs, Coptic, and Greek. This made it possible to work back from the known Greek and Coptic versions to an understanding of the Egyptian hieroglyphs.

1 *Egyptian Hieroglyphs* The translation of Egyptian hieroglyphs was only made possible by the discovery of the Rosetta stone in 1799 which gave the same edict in Greek, Egyptian demotic, and Egyptian hieroglyphs. Jean-Francois Champollion (1790–1832) was able to translate Egyptian hieroglyphs because the Rosetta stone gave the three versions of the same text. This gave Champollion access to the target language, which was a precursor form of Coptic. Prior to Champollion's discovery, it had been shown that the glyphs inside 'cartouches' represented the names of pharaohs. This was only possible because the target language was a human language which shared the (very) human concept of kingship with other languages. These discoveries allowed Champollion to develop the means of translating Egyptian hieroglyphs generally.

2 *Cuneiform* Cuneiform was deciphered over a period of about a century. Stages in the process included the discovery that the script was to be read in the direction of the wedges and that words were separated from each other by an oblique wedge. Next came the identification of the words for king and satrap. But progress accelerated in the early 19th century, when, in 1835, a British officer of the East India Company rediscovered the Behistun inscriptions in Persia which gave identical texts in the three official languages of King Darius' empire: Old Persian, Mesopotamian Aramaic, and Elamite. Thus, the Behistun inscriptions turned out to be the Rosetta stone for Cuneiform.

3 *Cretan Linear B* The Cretan Linear B was deciphered famously by Michael Ventris assisted by John Chadwick in the early 1950s. Here the key *was* the discovery that Linear B represented a form of Greek. Michael Ventris (1922–1956) succeeded in translating the Mycenaean Cretan script Linear B. In 1952, he discovered that it expressed an early form of Greek. This was the key to the translation. Ventris had thought previously that Linear B expressed Etruscan. John Chadwick had worked at Bletchley Park.

4 *Mayan Hieroglyphs* Similarly Mayan hieroglyphs were deciphered in a series of steps with the script only largely translated in the late 20th century. The key was the discovery that Mayan glyphs were related to syllables of modern Mayan languages. The Mayans were using their hieroglyphic script when they were conquered by the Spaniards in the 16th century and their meaning was lost. Mayan text was translated on the assumption which proved correct that it represented an ancient form of modern Mayan languages. The Russian scientist, Yuri Knorosov (1922–1999), discovered that Mayan glyphs were syllabic.

In these four cases, where lost scripts have been deciphered, it has only been possible by the fact that the underlying language has been known or where the lost language has fortuitously been associated with passages in a known language. Note also the importance of human concepts such as kingship, satraps, number (based on human limbs and digits), places, and dates.

Codes...

Enigma Code The famous deciphering of wartime German messages encoded by Enigma machines was a not a case of translation as the target language was German and known to be German. Any putative decryption *had* to make sense in German. Thus, it is not a good parallel for the decoding of alien messages. In the case of alien messages, the underlying language of the message is unknown and there is no means of discovering what it is. And the military context of the use of the Enigma machine was key in showing the shared military concepts of the cipherers and the de-cipherers.

... Untranslated Languages

Turning to scripts which have not yet been deciphered,

1 *Linear A* In the case of the Cretan Linear A, the script represents an unknown language. Like Linear B, Linear A was discovered by Sir Arthur Evans at the Knossos site in Crete. Archaeological evidence showed that it was a predecessor of Linear B. The language which it expresses has not been identified and it remains un-deciphered. It has though been possible to identify numbers. If the target language remains unknown, then it is unlikely to be deciphered.
2 *Rongorongo* Rongorongo, the script of Easter Island, has not yet been deciphered despite the target language being very likely Rapa Nui, an Eastern Polynesian language. Translation is hampered by there being only about two dozen texts on which to work. It is telling that even where the target language can be guessed at, the meaning of the script still remains opaque.
3 *The Voynich Manuscript* is a document in an unknown language which has been carbon-dated to between 1404 and 1438. The manuscript is on vellum and consists of about 240 pages and came to light in 1912. Since 1969, it has been held by the library of Yale University, and it is illustrated with pictures of people, plants (some fictitious), and astrological symbols. It has survived numerous attempts to translate it by cryptographers, both amateur and professional. All have failed. No progress has been made in identifying the language in which it was written. It is believed to represent the text of a natural language, but it is unclear whether it is a code, hoax, an artificial language, or just an unknown ordinary language. Suggestions have ranged between Nahuatl (the language of the Aztec empire), Old Turkic, and an 'unknown north German dialect'. Again, without knowledge of the target language, translation seems impossible.

All this is relevant to translating a message from an alien 'civilization'. Since we cannot know the language in which the original message was sent, the prospects for translation must be very limited. Further we will not have the advantage of shared concepts and practices. If we cannot decipher a language where we have shared concepts, how much more difficult will it be to translate a language where we have very few or more likely none? Further, there will be no Rosetta stone, giving a parallel text with a known language.

Any attempted translation would take place in the context of the stages in the formation of the alien message and its receipt and decryption by humans.

Alien 'language' > alien text > alien radio message > DECRYPTION > human language

The decryption (or translation) would reverse (or attempt to reverse) the first part of this process.

Alien radio message > alien text > alien 'language' > DECRYPTION > human language

In attempting to translate the alien message, first, the human translators would have to convert the radio message into the alien text. Then they would have to translate the alien text into the alien 'language'. Finally, they would have to attempt to translate the alien language derived into their preferred human language. This complex process is thus triply problematic.

First, to get from the alien message to the alien text, the human translators will have to discover the rules used to convert the original alien text into the alien radio message. And until these are known, they cannot discover the alien text on which the alien radio message was based. This would only be possible if they knew the alien language and could translate it into their preferred human language, as otherwise they could have no way of knowing whether their supposed knowledge of the conventions was correct. Trial and error can only succeed if you know what counts as error.

Second, if *per impossibile*, this were achieved, the human translators would then have the task of converting the alien text into the alien language. Again, the same difficulty would arise. Without knowledge of the alien language, the translators would have no criteria for successful conversion of the text into the alien language.

Third, if *per impossibile*, the second step were successfully achieved, the translators would then have to translate the alien language into a human language. But without parallel texts, with a language known to the human translators, this would be no more possible than current attempts to translate Linear A or Rongorongo.

It might be that the alien message was actually a direct transmission of alien speech, leaving out the alien text stage – like a *BBC Radio 4* broadcast. In such a case, the process would be as follows:

Alien radio message > alien 'language' > DECRYPTION > human language

But translation would be only a small advance on example (4), as the stage from alien text to alien language is omitted.

The translation project suffers from what might be termed 'cryptographic circularity' – or prosaically a 'chicken and egg' puzzle. We can only translate an alien radio message if we already know the alien language. But the alien language is what we are seeking to discover.

Science and Maths – Keys to Understanding Alien Messages?

One proposed solution to this problem of communication is that science and mathematics are universals which will be comprehensible to any extraterrestrials capable of interstellar communications. Thus Kathryn Denning, an anthropologist at York University Toronto, quoted in *Scientific American*, argued:

> Scientists – physical scientists and mathematicians in particular – tend to be more prone to thinking that because we'll be dealing with the same physical structures in the universe, we can use those as our Rosetta Stone, so to speak, and build up from there – send each other the value of pi, and then we're off to the races.
>
> *(Folger, 2011, p. 44)*

While we might not be able to understand alien culture, religion, law, or psychology because of the inevitable huge differences between humans and aliens, mathematics being objective and shared by any intelligent species, we could use mathematical truths as the common key to deciphering any message that we might receive from them. Mathematics would operate as a language shared by aliens and humans, and it could be used like the Greek on the Rosetta stone to decipher Egyptian hieroglyphs.

Similar assumptions about the universality of maths can be made about science. The laws of physics and chemistry must surely be the same throughout the universe, and it seems a reasonable assumption that their universal character could be used as the bases for understanding intelligent aliens who would have made the same discoveries as human scientists. The universal truths of maths might enable us to communicate with intelligent aliens and them with us. Prime numbers will surely have the same significance for aliens as they have for us.

Thus, SETI scientist Steven Dick argues that there could be 'overlaps' with alien science that will give us the much sought-after objectivity, and that acquaintance with alien science could give us 'objective knowledge'.

> If comparison is possible, one wonders whether the long-sought 'objective knowledge' might be found at last by gleaning the common elements remaining after processing by many sensory and mind systems independently evolved throughout the universe.
>
> *(2000, p. 47)*

Thus, it is claimed that science and maths could provide an answer to the question of how we could decipher an alien message, and how to send an easily understood message in the opposite direction. For example, in his novel, *Contact*, Carl Sagan imagines a powerful radio message from the star Vega which gave a list of prime numbers in binary form (1986/ 2022).

Maths and Science Expressed in Language[4]

But mathematics and science are parasitic on language. Only language users can do maths – and language has to be expressed in a natural language of which mathematical notation is an abbreviated and convenient form. What is more the meaning of these mathematical conventions is not obvious or 'objective'. They can only be explained by using a language already understood by the persons using the conventions. Ostension and counting out beans aloud, for example, could be used to explain number conventions, plus signs and so forth, to human children. But ostension would be out of the question in explaining human mathematical conventions to an alien and *vice versa*. For such conventions to work, the alien would have to already understand all the rules that underlie 'explanation', 'ostension', and 'counting out'. Further it would be necessary, but not a sufficient condition, for the explanation of number to an alien that he (she, it, or whatever) were physically present. In the case of radio messages, this, of course, would be impossible.

One way that we can learn a previously unknown language is by interaction with native speakers. This may involve (but may not be limited to) ostension. Famously, it was believed (incorrectly) that the name 'kangaroo' was the result of a misunderstanding of a question asked of a native and that it means 'I don't know'. In fact, the name derives from 'ganguru', meaning 'black kangaroo', the word for the eastern grey kangaroo in the speech of the Guugu Yimidhir people of north Queensland. Still this fable describes just the sort of mistake that could be made very easily. Note that in C. S. Lewis's novel, *Out of the Silent Planet*, the alien creatures are quite similar to human beings and their language is learnt by physically meeting and engaging with them (1938).

It might be argued, nonetheless, that numbers are sufficiently obvious for any race of intelligent aliens capable of transmitting radio messages that their use in a natural language could be easily understood. Surely, a list of prime numbers would be obvious in whatever language they were expressed. But even the expression of a list of significant numbers, such as the sequence of the primes or the expansion of pi, or other significant constants or equations would be just as impenetrable as any other list or expression. Suppose, for example, that the alien language expresses the first one million integers with discrete different symbols or that the language involves numerous different number systems for different purposes (as is the case in Japanese). Further, there is no reason to think that the alien language would use a familiar base, such as two (binary numbering), ten, or twenty. Beings built differently from ourselves might even have no need to use a number system which used bases for mathematical calculation. It would be an extraordinary coincidence if they used the same bases as we do.

And what is true of maths is also true of science. Both can only be expressed in language, and hence neither can be used as a Rosetta stone to the language in which they are expressed. In the case of an alien language, we would need to be able to understand it before we could extract any scientific or mathematical meaning. Hence, maths and science cannot operate as a Rosetta stone for the interpretation of an alien language.

Even if we could wish away the difficulty that alien science and maths can only be expressed in the alien language and hence cannot operate as a Rosetta stone, there is another overwhelming problem in using the supposed universals of science and maths to decipher an alien language. It might be thought that science is genuinely universal as humans, and aliens share the same universe subject to the same laws. Physics and chemistry will surely be the same in all parts of the universe.

Science is a human institution and is based on human interests, intellectual powers and abilities, sensory faculties, and on humanity's place within the universe. Aliens are likely to be in completely different circumstances and to have different faculties, powers, interests, and concepts. As a human institution, science depends on the existence of persons. In other words, only people and entities like people in a human or human-like society can do science. Further, aliens might be so constituted that they could see immediately the solution of *scientific problems which are for us extraordinarily difficult or wholly intractable*. If (and it is a monstrous if) alien science is like human science, then there can be no escape from interpretation and selection in the formation of scientific theories. Even if there is an overlap, it can no more give us 'objectivity' than if it were the possession of just one intelligent alien race.

That aliens could recognize mathematical and physical constants, or that they would have the same significance for them as they have for us, is far from evident. Attempts to develop an 'objective' scientific language based on mathematics is just as likely to be incomprehensible to aliens as a human language. And further, mathematics and other formal languages are parasitic on natural language. In other words, mathematics can only be expressed in a pre-existing natural language – it is only because we can talk (and write) that we can do maths. And we could only understand maths if we can understand the language in which it is expressed. Thus, the supposedly shared universals of maths and science could not act as a key to an alien language. A further serious difficulty, as pointed out by Nicholas Rescher, is that alien maths may be comparative rather than quantitative (1999, p. 199).

There is another reason why scientific theories could not provide some kind of shared objective truth. Wholly different scientific understandings could not account for the same phenomena. Some truths or ways of looking at the world may be possible to some intelligent species but may be impossible to others. As we have seen, aliens might have no difficulty in understanding the paradoxes of quantum mechanics or visualizing four-dimensional objects. Indeed, they would be at a loss to understand how we could be puzzled by quantum mechanics and its paradoxes. Their powers and abilities would lead to sciences that were very different from our own while dealing with the 'same' phenomena. *No two (or more) things are just the 'same' but are the same with respect to a particular characteristic.*

Of course, human scientific knowledge is no more or less lacking objectivity than any other human knowledge. Unless we accept a naive version of empiricism where 'data' are presented to us directly uninterpreted, objectivity is supplied by the process of investigation, experiment, and review by peers with the same methods, powers, shared understandings, and assumptions.

Pictures and Diagrams as 'Rosetta Stones'?

It might be thought that pictures could be the means of establishing a common frame of reference by which we could interpret an alien message. But similar difficulties would emerge. Pictures, images, and plans are expressed through a variety of different but overlapping conventions. Thus, different conventions govern pictures, such as Van Gogh's *The Church at Auvers* (where the ordinary rules of perspective are ignored) and blueprints for a steam engine or a diagram for a computer chip. It has been suggested that the inclusion of pictures and diagrams in an alien message might be a key to the translation of the message. If we did not know the convention used, we would have no means of knowing how to form a picture or a diagram from the message. Nor would we have any means of knowing that an image was being transmitted. Particular difficulties would be involved in the transmission of images involving multidimensional (3D or more) objects. The characterization of perspective and distance would present special difficulties. It is a purely human convention to represent nearby objects as larger than more distant ones. And how would we know that the image was the right way up?

Anti-cryptography

It has been suggested that an alien message might be designed to make it easy for us to translate it – 'anti-cryptography'. It follows that deciphering an alien message might be easier than the translation of Egyptian hieroglyphs of German wartime Enigma messages. This is because the alien message would be designed to be easily translated. As Clifford Pickover puts it, they would be practising "...*anticryptography*, the science of designing codes that are as easy as possible to decipher..." (1998, p. 149). The alien message would be in contrast to the German Enigma messages where the intention was to make deciphering as difficult as possible. Still, it is hard to see how aliens might do this without knowing any human language or linguistic conventions – or that we could recognize their attempt at anti-cryptography. It is of course an entirely human assumption (based on no evidence) that an alien civilization would do this, or would even want to do this. Or that the alien civilization would have the (very human) concepts that would make it even conceivable for them to do so.

Alien Syntax and Semantics

One possibility is the discovery that alien messages had syntax or internal structure. There might be evidence of iteration and self-reference. Expressions might be embedded in other expressions, for example. But although we could develop theories about the syntax employed in the alien message, we would have no clues as to what it meant – its semantics. Conceivably, it might have no substantive meaning for us, like a piece by Bach or an abstract painting – or whale song. We would have no means of telling whether the message had meaning or even what sort of meaning it had, if any. While alien syntax might be available to us, its semantics, if any, would always remain elusive.

Language has a multiplicity of uses. Chomsky has emphasized that communication in the strict sense – the transmission of factual information – is only one of a variety of roles that language plays. These include, for example, planning, ordering thought, persuading, deliberating, praying, teaching, etc. Wittgenstein made similar claims and argued that errors flow from attempting to assimilate different types of language use. Again, this suggests that we would be unable to identify the uses to which aliens were putting their language. And, in turn, this means that we would be unable to establish a common frame of reference by which we could discover the subject of an alien communication. Aliens could well have language uses of which we cannot conceive, and we would even have difficulty in identifying uses which were the same as or similar to our own.

The Channel Problem

Another difficulty in translating an alien message is what may be called the 'Channel problem'. All human languages are produced sequentially using sound (or sound equivalent) in a single channel. They can also be expressed in writing and its derivatives. But in all cases, it is in a single sequential series of expressions. These can be supplemented by signs and physical expressions – the use of hands and tone of voice, for example. But these are not essential to the expression and understanding of language.

But there is no reason why non-human languages should be so limited. Television usually involves two channels: sound and moving pictures, but often other channels are added: sign language, music, news, and 'tickers' covering stock market prices, exchange rates, and commodity prices. And notice that in television transmissions, both channels (sound and pictures) have to be properly co-ordinated for the message to be fully understood. Very often, any one channel is only fully comprehensible with the other. Alien languages (and messages) might employ multiple channels, and we might have no means of determining how they were related to each other. They might, for example, employ different channels for different parts of speech. And suppose we only received one channel with the other(s) unknown to us. While it may be easy enough for us to distinguish the different channels in human communications on screens, it would not be so easy to do the same for alien communications. Indeed, it would be a probably insoluble problem in its own right. And what if the alien message had 363 channels and we only had access to a few of them?

Conclusion

It is worth stating baldly that understanding any alien message is very likely impossible. Hopes of an exchange of messages appears equally forlorn. It may be possible for us to detect that a message has been sent, thus confirming the existence of aliens with the power to send messages. This might relieve the loneliness that Jill Tarter describes, but it would leave her and those like her in the tantalizing position of suspecting, or even believing, that aliens exist but wholly without the ability to understand their messages or to communicate with them. This would amount to a linguistic signature similar to (yet different in significance from) a techno- or bio-signature. In all likelihood, even a vast stream of alien messages would remain opaque to us forever without the possibility of successful interpretation. While we might discover intriguing facts about alien syntax, without a common frame of reference the meaning (if any) would always remain elusive. How important would it be? Here we might dwell on Peter Cook's remark: "As I looked out into the night sky, across all those infinite stars, it made me realize how insignificant they are."

Notes

1 Tarter, J., 2009. 'Join the SETI search', *TED: Ideas Change Everything*. www.ted.com/talks/jill_tarter_s_call_to_join_the_seti_search.html#1022000 (Accessed 12 September 2024).
2 The Fermi Paradox is named after the Italian physicist, Enrico Fermi, who asked why, if extraterrestrial life was so probable, there was no evidence of it.
3 Touch can be used by blind, deaf, and mutes. Helen Keller is the prime wonderful example. Berwick and Chomsky have pointed out that it is fortunate that humans cannot express language by smell (2016, p. 12).
4 This section benefited from Nicholas Rescher's discussion of alien science in *The Limits of Science*, pp. 197–222 (1999).

References

Baker, M. C., 2001. *The Atoms of Language*. New York: Basic Books.
Berwick, R. & Chomsky, N., 2016. *Why Only Us Language and Evolution*. Cambridge and London: MIT Press.

Bradbury, R. J., Cirkovic, M. M. & Dvorsky, G., 2011. 'Dysonian Approach to SETI: A Fruitful Middle Way'. *JBIS*, 94.

Cocconi, G. & Morrison, P., 1959. 'Searching for Interstellar Communication'. *Nature*, 184, pp. 844–846.

Davies, P. C. W., 1995. *Are We Alone? Philosophical Implications of the Discovery of Extraterrestrial Life*. London: Penguin.

Dick, S. J., 2000. 'Extraterrestrials and Objective Knowledge'. In Allen Tough (ed.), *When SETI Succeeds: The Impact of High-Information Contact*. Foundation for the Future.

Everett, D., 2005. 'Cultural Constraints on Grammar and Cognition'. *Current Anthropology*, 46(4), pp. 621–646.

Everett, D., 2013. *Language: The Cultural Tool*. London: Profile Books.

Everett, D., 2016. *Dark Matter of the Mind*. Chicago: Chicago University Press.

Everett, D., 2017. 'Chomsky, Wolfe and me'. *Aeon*. htttps://aeon.co/essays/why-language is not everything –that...ium=email&utm_terms=0_411a82e59d-cd9940f83d-69441609 (Accessed 11 January 2017).

Folger, T., 2011. 'Contact the Day After'. *Scientific American*, January, 304(1), pp. 40–45.

Jonas, D. & Jonas, D., 1976. *Other Senses Other Worlds*. New York: Stein and Day.

Koomin, E. V., 2012. *The Logic of Chance*. New Jersey: Pearson Education.

Lewis, C. S., 1938/1952. *Out of the Silent Planet*. London: Pan Books.

Maynard Smith, J. & Szathmary, E., 2000. *The Origins of Life*. Oxford: Oxford University Press.

Musso, M., Moro, A., Glauche, V., Rijntjes, M., Reichenbach, Buchel, C., & Weiller, C., 2003. 'Broca's Area and the Language Instinct'. *Nature Neuroscience*, 6(7), pp. 774–781.

Overbye, D., 2012. 'Search for Aliens Is on Again, but Next Quest Is Finding Money'. *New York Times*, 12 January. www.ndtv.com/world-news/search-for-aliens-is-on-again-but-next-quest-is-finding-money-568631 (Accessed 21 August 2022).

Pickover, C., 1998. *The Science of Aliens*. New York: Basic Books.

Pinker, S., 1994. *The Language Instinct*. Harmondsworth: Penguin.

Rescher, N., 1999. *The Limits of Science*. Pittsburgh: The University of Pittsburgh Press.

Sagan, C., 1986/2022. *Contact*. London: Orbit.

Tour, J. M., 2020. 'We Are Still Clueless about the Origin of Life'. In Thraxton, C. B. et al. (eds.), *The Mystery of Life's Origin*. Seattle: Discovery Institute Press, pp. 323–357.

Webb, S., 2015. *If the Universe Is Teeming with Aliens … Where Is Everybody?* Heidelberg: Springer.

Winch, P., 1964. 'Understanding a Primitive Society'. *American Philosophical Quarterly*, 1(4), pp. 307–324.

Wittgenstein, L., 1953/1968. *Philosophical Investigations* (Translated by GEM Anscombe) Oxford: Basil Blackwell.

Wittgenstein, L., 1980. *Culture and Value* (Translated by Peter Winch). Oxford: Basil Blackwell.

7

IF AN ALIEN COULD TALK, COULD WE UNDERSTAND IT?

David Ellis

Whether in sci-fi films, Reddit posts, or Congressional Hearings, there is a common intuition that if space-travelling aliens can visit Earth, then they will be able to learn any human language. Or, at the very least, that we will have enough mathematics and science in common to communicate. This chapter challenges these intuitions by examining Wittgenstein's views on language and meaning. We will trace how Wittgenstein's views change from a picture theory to a language game account of language and apply each to the question: if an alien could talk, could we understand it? We will see reasons for optimism and pessimism and propose that the above intuition should be exchanged with the following principle: whereof human and alien forms of life differ, thereof we cannot understand one another. The answer to whether we could understand a talking alien is not accountable to whether we're intelligent enough to decode their language, but whether we have enough in common to make sense of one another. We conclude that differences in forms of life lead to differences in meaning and that human and alien forms of life could be so different that complete mutual understanding is an unrealistic expectation.

Those familiar with Wittgenstein can likely guess from the title the direction this chapter is going to take. But for the sake of equal beginnings, we will briefly outline Wittgenstein's talking lion. According to Wittgenstein, although words are, by analogy, vehicles that carry meaning, we should not be tricked into thinking that words actually have meaning baked into them. Meaning is found in contextual use – words mean what they do and not what they are. However, words do not do things by themselves. We do things with them, and we can do whatever we please. So, meaning is more specifically what words are doing within a given context. Words are like pieces that are moved in games, and the pieces are moved following the game's rules; thus, Wittgenstein speaks of 'language games'. But just as how a word means what it does within a game, a game means what it does within the lives of those who play, and there are many forms of life. It follows that our understanding rests on our ability to contextualize what is said in the game(s) being played in the form of the life of the players. This explains Wittgenstein's remark "if a lion could talk, we could not understand him"; a lion's form of life is so different to ours that we cannot claim to know what the lion means by what the lion says (1994, p. IIxi; 223e). Put another way, we won't understand the lion not because we cannot speak English, but because the meaning of what is said is not

DOI: 10.4324/9781003440130-11

found in English. If this is the case with earthly lions, then what about otherworldly aliens? If an extraterrestrial intelligence (ETI) were to make contact, would we understand one another? Wittgenstein gives reasons for optimism and pessimism, and a great deal orbits around what we can hope or expect to have in common with aliens.

The main areas for optimism are the private language argument and ETIs (and humans) sharing family resemblances in our forms of life and language games. Regarding the first, although there are no inherently meaningful languages, there are no private languages either – that is, there are no languages that have their meaning logically closed off from everyone apart from the speaker. Consequently, any language (human or alien) can *in principle* be learned by any language speaker (human or alien). The challenge is whether the modality (e.g., visual, sounds, scents) of communication is accessible, and whether the speaker's form of life is like the interpreter's. This leads to the second area of optimism – family resemblance. We need a common form of life to share a common language, and common forms of life need not be identical. This is like how we recognize family members: each is not related by having the same quality, but by having various overlapping tendencies, mannerisms, appearances, and attitudes. All (or many) intelligent life forms may share common tendencies, customs, or circumstances that develop similar forms of life and language games. For example, if some of the ideas that aliens have are not innate, then they – like humans – might have something like 'learning through experience' and similar language games might develop to facilitate it, like describing what is sensed or asking and answering questions. It would be unreasonable to expect aliens to have universities and exam invigilators like we do, but it is reasonable to expect that if they have non-innate ideas, then they will have some kind of process for gaining them. But there is also pessimism in Wittgenstein's view due to the uncertainty of commonality and the difficulties that differences raise.

The problem is twofold: having similar biology, psychology, and conscious experiences does not automatically grant common meaning; and extreme differences in our biology, psychology, and conscious experiences could prevent common meaning from ever developing. Breaking this down, even if our lives were biologically, psychologically, and consciously alike, that itself does not give us common meaning. No matter what the aliens are like, they would not understand what we mean straight away because they do not understand what it is like to be human; they have never been a member of our form of life. For example, my friendship circle has jokes and phrases on account of our common history and experiences. The fact that you have the same biology and linguistic abilities as us does not grant you access to our friendship circle, and thus access to what we mean by what we say. But optimism creeps back in here. The fact that aliens are not currently members of our language speaking community does not mean they never can be. If they are like us in various ways, then we might eventually form bonds and come to make sense of one another. After all, we are as much their aliens as they are ours, and the more alike we are, the more likely we will encounter similar issues. If we are one another's first encounter, then our coming to work through these issues might be the best bonding experience we could ask for. So perhaps we could join one another's communities or develop a new one sufficient for language. But that would be no easy task because of the second factor: aliens might be so different to us that we would struggle to ever develop a common form of life and language. Humans and aliens need to have things in common to have similar forms of life that allow for a common language to develop, but we might be so different in terms of biology, psychology, and conscious experience that the distance between us is simply too far to be bridged or for common ground to be found. We might have different senses, emotions, and psychological processes; our social-cultural context is likely different; our conscious experiences could wildly differ in terms of memories, imagination, and perceptions of time; our values, motivations, and thoughts about life, meaning, and purpose are again likely to differ. Aliens might understand and sense the world in ways we cannot; have ambitions, goals, and emotions that are incomprehensible to us; or even have no such desires whatsoever. More than underscoring how easily an alien's form of life could differ from ours, these factors suggest that the meaning of some regions of alien discourse lies outside our accessibility and understanding. So even if they visit Earth, they might never visit our world. They will bring their world with them and be unable to leave it.

Wittgenstein, therefore, gives reasons for optimism and pessimism. To see this in greater detail, we will examine how Wittgenstein's picture theory and language game account apply to human–alien communication. We will then explore how alien an alien's form of life could be, and how this undermines the intuition that we could communicate through maths and science.

Picturing Talking Lions

Humans have two bad habits when it comes to thinking about language. First, we tend to assume that languages only differ in their words and patterns; second, we also tend to assume that while words and patterns differ between languages, what they mean remains the same. We understand 'good morning' (in English) and 'bora da' (in Welsh) as two ways of meaning the same thing. This leads to the view that we would understand an alien language, given enough time and resources, because they will mean what we mean and it's just a matter of deciphering their symbols and patterns. We also intuit that if aliens are advanced enough to travel to Earth, then their resources will be so plentiful that they will be able to save us the task. Wittgenstein's later views find that to be a muddled way of thinking about language, but his earlier views are more sympathetic.

Wittgenstein's philosophy is usually divided into two periods: the early period of the *Tractatus Logico-Philosophicus* (*Tractatus*) and the late period of the *Philosophical Investigations*. His earlier work is more sympathetic to the intuitions this chapter challenges because the *Tractatus*, Churchill explains, "combines atomism with the picture theory of the proposition…", where there,

> are names which correlate with objects. Objects, then, on the ontological side of this equation, must be the simple constituents of facts, the elementary, unanalyzable building blocks of facts. As propositions are made up of names standing in relation to each other, facts are made up of objects in relation. What these relations are in propositions and in facts, and how they correlate, make up the picture theory.
>
> *(1994, pp. 391–392)*

The picture theory divides things into two sides: on the ontological side, 'the world is the totality of facts, not of things', and facts consist of simpler atomic facts and each atomic fact is composed of objects and their relations (Wittgenstein, 2022, §1.1). On the language side, names stand for objects and can relate to one another to form propositions, and in this sense, language pictures the world by showing us what the world is like. But the world is complex and changing, and so propositions need to come together to give a living picture of the living world. Propositions do this where "one name stands for one thing, and another for another thing, and they are connected together. And so the whole, like a living picture, presents the atomic fact" (Wittgenstein, 2022, §4.0311). A living picture is like when a witness uses figurines in court to show a jury what they saw. Names, like figurines, stand for objects; how the names relate, like how the figurines are ordered, shows their relations; and when seen together, a picture is given, and our task, like a jury, is to judge whether the world resembles the living picture. Picturing, therefore, does not require a person to perceive a mental image, instead, the term only means to bring the *action* of picturing to mind (Anscombe, 1971, pp. 18–19). In this view, all languages are meaningful because they picture through propositions but as different languages have different words and grammatical rules, not all languages picture in the same way.

There needs to be an explanation for how different languages are meaningful for the same reason; there must be a general form of a proposition that can be expressed in different ways by different languages. Consider how 'it is raining' and 'raining, it is' are different sentences that say the same thing (they express the same proposition). Yet, we utter different propositions when I say, 'I am hungry' and you say, 'I am hungry'. If we knew what the general form of a proposition is, then we would know something essential for every language, including alien ones. Here Wittgenstein answers, "The general form of a proposition is: This is how things stand" (2022, §4.5). Learning an alien language should be reducible to learning the alien's technique of picturing, which is not only how the alien draws a picture (in terms of their words, grammatical patterns, gestures etc.) but how aliens use pictures (what does comparison look like when an alien does it?). This resembles the intuition that the core challenge to human–alien communication is deciphering the messages. When we compare propositions to facts, we are looking to see if they match up – whether the propositions show us how the world factually is. This causes the correspondence theory of truth to be a natural bedfellow for the picture theory.

The correspondence theory of truth holds that truth is a matter of correspondence between what a claim says and what the claim is about, but variations differ in terms of strictness. A strict version, such as held by Russell or Moore, holds that as propositions are relations of names which stand for relations that constitute facts, then a proposition is true if and only if every name and relation perfectly correspond to reality (Lacey, 1996, p. 358; Baldwin, 2001, pp. 37–38). This is a high bar and arguably inapplicable to our ordinary use of language as we tend to complain when a person scrutinizes the meaning of every word we say. Looser versions do not require each name and relation to perfectly correspond but rather request that enough correlation occurs alongside interpretative freedom to find the proposition agreeable or disagreeable. To illustrate, a strict version would take 'I am going to bed' to be true *if and only if* I am literally, immediately, and directly travelling to my bed. Looser versions allow for some (welcomed) interpretative freedom. How might this apply to human–alien communication?

The social-cultural worlds that humans and aliens have are likely to be different, but the world of facts is likely to be the same. Combined with the view that all languages, including alien ones, are meaningful due to picturing the world, then there should be similarities between how human and alien languages work. Adding to the optimism for mutual understanding, it is notable that linguists are incredibly skilled at analyzing and learning an extraordinarily diverse range of human languages, and these skills will be helpful when it comes to human–alien communication (Vakoch and Punske, 2023). Under the picture theory, human–alien communication likely faces two overlapping challenges even if the world of facts is the same and our languages are similar. First, whether we can recognize one another's communicative modalities (i.e., the sensory mediums used to express language), and second, whether we conceptualize the world in recognizable ways (i.e., that our conceptualizations are not so different as to render us without any common ground). Call the first the *modality concern* as it concerns detecting, analyzing, and reproducing the modality of language; the second is the *conceptual concern* as it concerns comprehending, relating to, or imagining the subject matters being spoken about. This second challenge asks whether we can understand the facts as the *alien* understands them: facts might be universal, but our comprehension of them may diverge to such an extent that we can never 'see' or 'get' what the other means.

These two concerns overlap when a person uses sign language to explain what colours look like to the blind. The blind can neither *see* visual modalities of communication like hand movements nor *conceptualize* colours as the non-blind can. Yet we can nonetheless engage in meaningful conversations because we still have a great deal in common, including other senses. If humans and aliens share at least one sense in common (or have technology that can do the detecting for us), then we should eventually recognize the modality of one another's communicative acts. However, recognizing the modality of an alien's language does not mean that we will understand what is being said, just like how shouting about colours to the blind will still fall on deaf ears. There are, however, ways of circumnavigating these sensory differences which might be useful with aliens. The blind cannot make sense of what shapes *look* like, but those who can see share other senses with those who cannot and that suffices for both a common modality of communication (like vocal speech) and a common sensory conceptual framework (like touch) for communication to occur. This lets us place objects into the hands of the blind and vocally associate terms that stand for their visual appearances with terms that stand for their texture. Over time, the blind map visual language onto touch and grasp what is being said about the geometric properties of objects. There is, therefore, room for some optimism in this regard. If humans and aliens share a common world of facts (we seem to), and if all languages are meaningful by picturing those facts (as Wittgenstein suggests), then we should – given enough time and resources – be able to learn one another's languages, even if we have different senses. Having now outlined the picture theory of language as it relates to human–alien communication, let us move to Wittgenstein's language game account.

The *Investigations* is a repudiation, but not outright renunciation, of the *Tractatus*. Wittgenstein described his earlier work as mistaken but not entirely wrong: "it was not like a bag of junk professing to be a clock, but like a clock that did not tell you the right time" (Anscombe, 1971, p. 78). He believed that the *Tractatus* was on the right track, but the picture theory missed the mark in its inability to explain the meaning of sentences that do not picture. If language is meaningful because it pictures and pictures are always truth-apt, then non-truth-apt sentences must not picture and must therefore be meaningless. Yet, language does not consist of just declarative sentences; there are imperatives, questions, remarks, and other sorts of utterances that are not entirely in the business of reporting facts. Furthermore, these kinds of utterances are essential to key areas of human life like education, law, friendship, and politics. This leaves a substantial gap in what Wittgenstein can account for, and he took it to indicate that he was going about things the wrong way.

In his search for a common essence to language, he lost sight of how language is commonly used. Picturing, he concluded, is not essential to language. It is one, albeit common, language activity.

Wittgenstein went on to look at how language is used in everyday life and noticed two things: speaking is like playing a game where the things we say (or the moves we make) are meaningful because of their function in the language game, and the language games we play are meaningful because of their association with the speaker's form of life. Wittgenstein captures this link, remarking "to imagine a language means to imagine a form of life" (Wittgenstein, 1994, §19). We need to understand the speaker's form of life to understand what it means for them to play a game and what it means for them to say things within that game. As aliens do not share our form of life then they, like lions, cannot mean what we mean. But there is room for optimism. There are no languages that have their meaning logically closed off from us, and aliens might have similar forms of life and language games. Let us look at each in that order.

Wittgenstein asks us to imagine a language where "the individual words of this language are to refer to what can only be known to the person speaking; to his immediate private sensations. So another person cannot understand the language" (Wittgenstein, 1994, §243). Wittgenstein is not asking us to imagine a forgotten language or a heavily encrypted code. He is asking us to imagine a language that has its meaning logically closed off from everyone other than the speaker because the meaning is accountable to the speaker's private sensations. This is a *private* language, not because we cannot hear the speaker's voice but because we cannot access its meaning. Wittgenstein concludes that such languages are impossible for several reasons that exceed the scope of this chapter to explain, but it is worth noting the following. If mutual understanding depends on participants accessing what is being spoken about, then we would be unable to meaningfully talk about our feelings, ambitions, pains, and dreams – yet we obviously can. Indeed, if language is used for communicating, then the meaning cannot be private because there would be no meaningful communication if it were. So even an alien's language would not be private, and so it must be publicly accessible, understandable, and learnable for humans (and vice versa concerning human languages for aliens) (Garver, 1990, p. 193). We must, however, be careful to stipulate that the impossibility of any private language does not guarantee that we will learn any alien language. Public meaning can be so difficult to access that, in practice, it may as well be private. But ideally, there will be similarities in the activities that intelligent creatures, like humans and ETIs, perform.

In moving away from the picture theory of language that carried the general form of a proposition as baggage, Wittgenstein also moved away from the idea that every language will share one essential thing in common (Wittgenstein, 1994, §65, §67). He explains that

> instead of pointing out something common to all that we call language, I'm saying that these phenomena have no one thing in common in virtue of which we use the same word for all – but there are many different kinds of *affinity* between them.

and this is like the affinity that holds family members together (Wittgenstein, 1994, §65). No one thing unites the members of a family, but various similarities and tendencies do due to common origins, experiences, and upbringings. In this view, language games can share family resemblances and can be organized under family names like 'poetry', 'science', and 'teaching'. If humans and aliens share common experiences and upbringings, then we might develop similar forms of life and language games. Of course, this does not mean that an alien language will be expressed through sounds or that their linguistic customs will resemble ours. Instead, it means to say that aliens might have some things in common with us, and they might use language in similar ways as we do in those respects. Aliens, like humans, might have memories of past events and use language to talk about them; they might have emotions which are expressed through remarks; they might need to convey information to their young in a way that resembles teaching. What their memories are like, what emotions they have, and what information they convey may be very different to what humans are accustomed to, but using language in those contexts for those reasons is not so different.

In this section, we applied Wittgenstein's picture theory and language game account to human–alien communication and identified areas for optimism and pessimism. If the picture theory holds, then we can expect alien languages to share the general form of a proposition which opens paths towards mutual understanding. However, the picture theory suffers from its inability to account for the meaning of non-propositional linguistic actions like questions, commands, and remarks. These linguistic actions are essential for many activities performed in our – and perhaps alien – forms of life, and so being unable to account for them translates into a broader problem of being unable to apply to ordinary human (or alien) life. Wittgenstein's language game account holds that mutual understanding depends on one's familiarity with the language game and the speaker's form of life. So, whether we could understand an alien is answerable to whether we have similar forms of life or whether similar forms of life could develop over time. We identified two areas for optimism in this respect. First, the impossibility of private language means that every human/alien language is learnable for any human/alien; second, intelligent life forms might share family resemblances in their forms of life and language games. In the next section, we examine whether an alien's form of life could be so different that it prevents mutual understanding from occurring.

What Is It Like to Be an Alien?

In his essay *What is it like to be a bat?* Thomas Nagel writes,

> Conscious experience is a widespread phenomenon... No doubt it occurs in countless forms totally unimaginable to us, on other planets in other solar systems throughout the universe. But no matter how the form may vary, the fact that an organism has conscious experience *at all* means, basically, that there is something it is like to *be* that organism.
>
> *(1974, p. 436)*

Nagel is making three points: conscious experience is not exclusive to humans and is probably found throughout the Universe; there are different forms of conscious experience; and when an organism has conscious experience, then there is something which it is like to *be* that organism. Nagel uses the phrase 'subjective character of experience' to refer to what it is like for an organism to be that organism, and as indicated by the phrase, the character is seen from a subjective perspective and takes a form relative to each subject. Bats, he writes, are likely conscious and so there is something which it is like to *be* one. We also know that bats do not rely on vision as humans do, and instead perceive the world via echolocation that allows them to navigate through pitch-black environments at considerable speed. Nagel wants to know what the subjective character of that experience is like – what is it like to be a bat?

We could try guessing, of course. We could imagine what it would be like if our consciousness was transported into a bat and we lived the rest of our lives dangling upside down, having wings and chasing moths at dusk. That would certainly be an alien experience, but it would only tell us what it would be like for *us* to be a bat. Nagel wants to know what it is like for a *bat* to be a bat. It is not enough to consider in *objective* terms how different a bat's conscious experience might be to a human's (e.g., having echolocation or not), we must also consider in *subjective* terms the difference in what it is like to *be* a bat or a human. This second consideration is highlighted by distinguishing what it would be like for a human to wake up in the body of a bat versus a bat waking up in the body of a bat. This also echoes Wittgenstein's warning about not being deceived by a talking lion; the lion speaks English, but the lion is not an Englishman in lion clothes. Combining Nagel's bat and Wittgenstein's lion highlights a unique relationship that applies to human–alien communication: if what it is like to be a bat involves echolocation and that influences a bat's form of life, then differences in biology, psychology, or consciousness can result in differences in forms of life and language. This would be just as true about aliens as it is about bats, lions, and humans.

It is difficult for a non-bat to imagine what it is like for a bat to be a bat, but it is easy to imagine how what it is like to be a bat could differ from what it is like to be a human in ways that would impact language. Nagel, after all, shows in relatively clear ways how a bat's subjective character of experience could differ from ours because of them having an echolocation sense that we do not have. This would be an important factor when it comes to our inability to understand the bat if it were to say that a tree looks beautiful. Not only do we not see objects in echolocation form (so we don't know what trees look like from the perspective of a bat), but we don't know what it would be like for an echolocation object to appear beautiful. In effect, it is already difficult to explain what red or blue looks like to a blind person, never mind how red looks dangerous and blue looks calming. We can also imagine how our form of life and language would change if certain aspects of our conscious experience were to change. For example, imagine the consequences to notions of privacy, aesthetics, and trustworthiness if humans were to wake up with the ability to see in the ultraviolet spectrum or have a truly photographic memory. This is relevant for human–alien communication because meaning is contextual to one's form of life, and differences in biology, psychology, or consciousness can lead to differences in forms of life and thus differences in meaning. It is imaginable that aliens, like bats, have quite different forms of conscious experience than humans. But what if ETI are not conscious, or if they are, that their consciousness is not like ours?

ETI could take the form of artificial intelligence, p-zombies, or having something it is like to be them which is unlike what it is like to be us. Future AI might be capable of taking care of itself if a disease were to lead to human extinction, and this might have already happened on other planets with aliens. Maybe most organisms that can create interstellar vehicles are unable to survive the journey, so we should expect to find ETI-AI drones in the sky before we encounter their organic creators. Alternatively, ETI might be p-zombies (philosophical zombies) which are hypothetical entities that act just as humans do despite having nothing it is like to be one. They are zombies that are more interested in convincing us that they are 'real people' than they are in eating our brains. Consciousness might be a quirk of evolution and most intelligent life could be a zombie-like variety. Consciousness might, however, be as common as Nagel suggests, but it might come in a diverse range of forms. Some organisms might have a subjective character of experience that is so unlike ours that no human paradigm can express what it is like for them to be them. There might be aliens with no 'ghost in a machine' sensation; some might be confused by free will as they experience reality in an entirely deterministic way; and where humans typically identify as a single person within a single body, aliens might take their bodies to harbour several people. In any case, non-conscious forms of ETI like p-zombies or AI are conceivable, and so too are ETI with forms of consciousness inexpressible through human paradigms. These are relevant to one's form of life and the language games played in it, and future research should address this in greater detail than I can in this chapter.

We asked what the subjective character of echolocation is like from the perspective of a bat and answered that we cannot know because we do not have experiences like that, and even if we did, that would only tell us what it is like for a *human* to have it. Differences in what it is like might hinder some communication, but it need not result in a total inability to communicate. We earlier saw ways to circumnavigate modality and conceptual concerns when speakers have different senses, perhaps something similar can happen with different forms of conscious experience. If there are mind-independent facts accessible through conceptual frameworks like mathematics, science, and logic, then humans and aliens might have common forms of life and language in those respects, irrespective of what form our consciousness takes. But there is some room to argue that we shouldn't be surprised to find an alien's mathematics, science, and logic to be as alien as their fashion.

We expect mathematics, science, and logic to offer a basis for communication because we think those sorts of facts are mind-independent and will take the same form in alien conceptual frameworks. It seems that mathematical principles and facts about chemistry are true of the world and do not depend upon the perspectives, opinions, beliefs, or attitudes of any given mind. We also think that the conceptual frameworks that lead to such knowledge will work for all rational intelligent animals. Of course, there will be differences but no matter how aliens phrase it, they will agree with what *we* mean by $a^2 + b^2 = c^2$. There are pushbacks against each of these, the first being that laws of logic are laws of thought, and the second being that aliens could have alien forms of science.

Frege first imagines a psychologistic logician who thinks that the laws of logic are relative to our psychology, and then objects that the logician "conflates the laws of psychology (the laws of takings-to-be-true) with the laws of logic (the laws of truth)" and ends up being unable to distinguish the subjective from the objective (Conant, 1992, p. 142). In the psychologistic logician's view, the law of identity should be phrased: "it is *impossible for beings like us* (with the relevant population appropriately circumscribed) to acknowledge an object to be different from itself" (Conant, 1992, p. 143). According to the psychologistic logician, aliens with very different psychologies could have very different logical principles. If the alien's laws of logic are as foundational to their scientific and mathematical conceptual frameworks as our laws are to ours, then aliens likely have forms of science and maths that seem illogical to us. This undermines the intuition that humans and aliens will be able to communicate through common scientific and mathematical principles. But Frege counters the psychologistic logician's position by pointing out that if they understand the question "whose inferences are correct, ours or theirs (or neither)?" then they would be endorsing logic as something outside of psychology (Conant, 1992, p. 144).

When this logician says that the laws of logic are relative to psychology, they appear to be reporting something which they take to be objectively true about logic and psychology. But if they are right, and the laws are not objective laws of truth but relative laws of thought, then there is no objective standard of truth for the logician to appeal to if we were to disagree with them. All this logician could be saying when they say, 'the laws of logic are psychological' is 'those with psychologies *like mine* cannot avoid thinking about logic in this way'. Hence by understanding our question, they would be showing us, as much as themselves, that they *do* understand the laws of logic as laws of truth. Even by the logician's standards, no thinking thing can think beyond its capacity (and the laws of logic are, for this logician, that capacity), and not all thinking things share the same capacity. I might have different psychological capacities to the logician and consequently form a different understanding of logic. My belief that the laws of logic are not relative to psychology could be that understanding. That is to say, the belief that 'logic is not relative to psychology' could be an unavoidable output of how my psychology happens to work. The logician would not be able to say that my belief is false without undermining themselves by appealing to a logical basis of truth which is independent of psychology. Whether or not we agree with Frege, this is nonetheless a good example of how our most engrained intuitions about reason could deceive us when it comes to thinking about human–alien communication. At the very least, the topic is not as settled or clear-cut as our intuitions might suggest. Quine and Putnam further show that we could encounter aliens with conceptual frameworks that do not resemble ours even if the laws of logic are mind-independent.

According to Quine, there are right and wrong answers in science because there is a matter-of-fact reality to which scientific beliefs correspond, but there is no such reality to the meaning of language and so there is no objective standard to judge a translation's accuracy against (Føllesdal, 2013, p. xxi). In a Wittgensteinian sense, the activities performed in scientific experiments, along with their theories and conclusions, are inseparable from our psychology and society (Bolton, 1979, p. 343). They are not just how we understand the natural world, but are expressions of our social-cultural world, and this causes them along with the language games that take place within them to have a familiar appearance. This resembles Quine's view that we cannot translate "'Neutrinos lack mass' into the jungle language" – the speakers of the jungle language neither know nor participate in science, so there is no appropriate context to situate the sentence (2013, p. 69). The issue is Wittgensteinian in this respect; language differs where forms of life differ, and language is absent where context is absent. The reason why talk about neutrinos does not translate into 'jungle language' is not simply because there are no equivalent words for 'neutrino' or 'sub-atomic'. It is because there is no conceptual framework which could situate the meaning of those words or those sorts of sentences. We cannot pick up any sentence we please and translate it into any language we like because not all forms of life share the same customs or practices that facilitate such language games. Therefore, "if a culture is so remote from our own that one of our practices is altogether absent from it (or vice versa), translation of the language used in that practice could hardly be attempted" (Bolton, 1979, p. 343). If the alien's form of life is such that their scientific method has practices ours lack or lacks practices that are essential to ours, then there will be regions of scientific discourse where translation might be a fool's errand. But we should not think that a society either has science as we know it or none whatsoever. Even if we reject the psychologistic logician's view for Fregean reasons, we can still sympathize with the general thrust of the point. An alien's psychology may differ from ours and lead to further differences in how we each perform science, even if these differences do not involve logic. For example, there could be aliens that do not have a psychological habit of expecting future events to resemble past ones and therefore develop a form of science where the problem of induction never arises.

Induction is a method of inferring a general position from individual instances; each swan I have seen has been white, so I infer that the next swan I will see will also be white. The problem is that, contrary to intuition, "it implies no contradiction that the course of nature may change, and that an object seemingly like those which we have experienced, may be attended with different or contrary effects" (Hume, 1998, 4.2.18). The problem is multifaceted: knowing that inductive reasoning is problematic does not prevent us from depending on it, and inducing future events based on past ones is a core feature of science. If there was a unifying principle that connects individual events in a way that provides grounds for belief that the future will resemble the past, then inductive reasoning would avoid this problem. If, for example, causation was shown to exist and that fire always causes smoke, then we would be justified in our expectation to find smoke after seeing a fire. Hume famously expresses doubt about whether causation can be known to exist as a feature of the world as opposed to a tendency of how our minds process information, again emphasizing for our interests the impact that differences in psychology can have on differences in conceptual frameworks. My point is not that inductive reasoning is a problem nor that causation does not exist, but that there are philosophically respectable reasons to associate an individual's psychology with their conceptual frameworks. Aliens might have such a different psychology that they do not induce as we do; they might struggle to understand 'causation'; or they might see unifying principles everywhere. In short, we should not assume that an alien's science will work like ours just because we are intelligent animals in the same universe that features the same facts.

Putnam also thinks that the psychologistic logician misunderstands logic because their position is self-refuting, and he goes on to describe what it would take for him to change his mind about logic. He explains that a fundamental logical law will be 'necessarily necessary', that is, not necessarily true by chance or some contingent reason, but necessarily true by *necessity* (Conant, 1992, p. 124). However, that does not mean we will never change our minds about logic. Although logic has in a sense remained the same since Aristotle, our understanding of it has changed. What would cause us to change our minds about logic and how does this relate to aliens? Putnam explains:

> The idea is that the laws of logic are so central to our thinking that they define what a rational argument is. This may not show that we could never change our mind about the laws of logic, i.e., that no causal process could lead us to vocalize or believe different statements; but it does show that we could not be brought to change our minds *by a rational argument...* [The laws of logic are] presupposed by so much of the activity of argument itself that it is no wonder that we cannot envisage their being overthrown... by rational argument.
>
> *(1996, pp. 109–110)*

Although rational arguments will (probably) change our minds about some laws of logic and their characteristics in the future, a dramatic change, like the overturning of the law of identity, will not come from a rational argument which depends on the law of identity. We require something outside those parameters – we need to see another way of seeing logic – and that might happen if we encounter aliens. This, though, assumes that if aliens have 'alien logic', then we would recognize it as logic. If the laws of logic exist outside human psychology, then it is conceivable that some laws of logic are beyond human (but not alien) psychological access or comprehension. Humans might be hardwired to psychologically process information through certain principles that map onto objective laws of logic, and yet there could still be laws of logic that lie outside our programming. The point here is not that this is certainly true, but that it is conceivable, relevant to human–alien communication, and challenges the intuitions we outlined at the start of this chapter. If an alien had an 'alien logic', there is no assurance that we would comprehend it and their science might work for equally incomprehensible reasons.

Conclusion

Wittgenstein gives reasons for optimism and pessimism depending on whether we endorse either of his accounts. On the picture theory, human and alien languages will share commonalities because all languages convey meaning by picturing the world through propositions, and the world, as the totality of facts, is the same for humans and aliens. The main challenges here involve whether we have compatible communicational modalities and conceptualizations that can be mutually understood. Optimistically, commonalities in language and facts should be a good basis for building common modalities and conceptualizations. On the pessimistic side, it is not obvious that humans and aliens will have sufficiently similar senses, psychology, or conscious experiences to develop mutually accessible and understandable modalities and conceptual frameworks. There is also the overall weakness of the picture theory's inability to account for non-picturing utterances. The language game account also offers optimism and pessimism. On the one hand, all languages are (in principle) understandable because there are no private languages, and there might be a family resemblance between humans and aliens seen in the activities and language games which feature in our forms of life. On the other hand, differences in forms of life are differences in domains of meaning, and we would not understand aliens any more than lions. Moreover, the biological, psychological, and conscious differences between our life forms might be so extensive that we will struggle to ever develop a common form of life sufficient for a common language.

So, if an alien could talk, could we understand it? The Wittgensteinian answer is 'whereof human and alien forms of life differ, thereof we cannot understand one another'. The consequences are that in almost every first-encounter scenario, we will not understand one another because we do not share a common form of life. Moreover, our biological, psychological, and conscious differences may prove to be incredibly challenging barriers to overcome for a common form of life and language to develop. We would either need a lot of stars to align just right for us to have (or be able to have) a common form of life and language, or for an independent principle, like evolution, to cause intelligent life forms to develop similar forms of life. The most optimistic view is that a combination of evolution and principles of reason will cause intelligent animals to develop similar forms of life, practices, and language games. The most pessimistic view is that aliens and humans will be so different that despite recognizing one another's intelligence, we will be unable to adopt one another's point of view for mutual understanding to occur. I am currently inclined to hold a pessimistic, rather than an optimistic, view, but I hope that the other chapters in this book will change that.

References

Anscombe, G.E.M. (1971) *An introduction to Wittgenstein's Tractatus*. 4th ed. London: Hutchinson University Library.

Baldwin, T. (2001) 'Bertrand Russell (1872–1970)', in A. Martinich and D. Sosa (eds.), *A companion to analytic philosophy*. Massachusetts & Oxford: Blackwell Publishers (Blackwell Companions to Philosophy), pp. 21–44.

Bolton, D.E. (1979) 'Quine on meaning and translation', *Philosophy*, 54(209), pp. 329–346.

Churchill, J. (1994) 'Wonder and the end of explanation: Wittgenstein and religious sensibility', *Philosophical Investigations*, 17(2), pp. 388–416. Available at: https://doi.org/10.1111/j.1467-9205.1994.tb00106.x

Conant, J. (1992) 'The search for logically alien thought: Descartes, Kant, Frege, and the Tractatus', *Philosophical Topics*, 20(1), pp. 115–180.

Føllesdal, D. (2013) 'Preface to the New Edition' (2013) in Quine, W. V. (ed.), *Word and Object*. New ed. Cambridge, Mass: MIT Press, pp. xv–xxviii.

Garver, N. (1990) 'Form of life in Wittgenstein's later work', *Dialectica*, 44(1/2), pp. 175–201.

Hume, D. (1998) *Enquiries concerning human understanding and concerning the principles of morals*. 3rd ed., 15. impression; reprinted from the posthumous ed. of 1777. Edited by L.A. Selby-Bigge and P.H. Nidditch. Oxford: Clarendon Press.

Lacey, A.R. (1996) *A dictionary of philosophy*. 3rd ed. London and New York: Routledge.

Nagel, T. (1974) 'What is it like to be a bat?', *The Philosophical Review*, 83(4), p. 435. Available at: https://doi.org/10.2307/2183914

Putnam, H. (1996) *Philosophical papers, volume 3: Realism and reason*. Cambridge: Cambridge University Press.

Quine, W.V. (2013) *Word and object*. New ed. Cambridge: MIT Press.

Vakoch, D.A. and Punske, J. (eds.) (2023) *Xenolinguistics: Towards a science of extraterrestrial language*. Abingdon and New York: Routledge.

Wittgenstein, L. (1994) *Philosophical investigations*. 3rd ed. Translated by G.E.M. Anscombe. Oxford: Blackwell.

Wittgenstein, L. (2022) *Tractatus Logico-Philosophicus*. Side-by-side ed. Translated by C.K. Ogden et al. London: Kegan Paul.

PART IV
Religion

8

DHARMIC PERSPECTIVES ON EXOPHILOSOPHY

The Hindu, Buddhist, and Jain Traditions on the Possibility of Extraterrestrial Life

Jeffery D. Long

Introduction

This chapter will explore how the Hindu, Buddhist and Jain traditions would respond to the discovery of extraterrestrial life by examining the religious texts of these traditions as well as the views of various modern teachers within them who considered this possibility. There are many reasons to expect not only that the eventual discovery of and encounter with extraterrestrial life would *not* present a major religious problem for these traditions, but that such a discovery could be seen as confirmation of elements of the worldviews of each. These traditions will be explored in tandem because of the great overlap among them in terms of cosmology, particularly regarding issues relevant to the search for extraterrestrial intelligence. All three affirm a vision of the cosmos as vast and ancient, on scales comparable to those used in modern science. In the literatures of all three, non-human intelligences play roles comparable to those of extraterrestrial life forms in contemporary science fiction. All three describe technologies that are also compatible with the idea of intelligent beings capable of traveling to the Earth from other worlds, though this chapter will *not* affirm the popular "ancient alien" hypothesis that is sometimes invoked to explain the presence of such technologies in premodern texts (and reasons will be given for skepticism in this regard). Finally, all three traditions describe worlds in the cosmos, other than the Earth, which are inhabited by intelligent beings. While in many cases it would be better to interpret some of these references to other worlds as other universes or planes of existence, rather than planets in the conventional sense, again, the idea that there are places other than the Earth that are home to intelligent life forms is not a foreign concept from the point of view of these traditions.

Regarding potential extraterrestrial religious beliefs and practices, this chapter will draw attention to the emphasis these traditions have laid upon inclusivism and pluralism in approaching views other than their own. It will suggest a *cosmic pluralism*, rooted primarily in the teaching of the Hindu master, Swami Vivekananda, but also in comparable Buddhist and Jain concepts, such as *upaya kaushalya* (the Buddha's "skillful means" in teaching diverse audiences) and *anekantavada*, the Jain doctrine of the complexity of existence and the consequent multiple valid points of view from which truth can be approached.

DOI: 10.4324/9781003440130-13

In short, this chapter will argue that there are good reasons for Hindus, Buddhists, and Jains to see the discovery of extraterrestrial life as a welcome development, and to see these three traditions as potential resources for human adaptation to the reality of extraterrestrial life, whenever such life is established to exist and is even encountered.

Religion and the Question of Extraterrestrial Life: The Case of Christianity

There are good reasons, from within their respective traditions, for Hindus, Buddhists, and Jains to see the potential discovery of extraterrestrial life as a most welcome development. Indeed, these traditions have been affirming the existence of worlds other than the Earth and non-human intelligences for millennia. It is possible, therefore, to see these traditions as potential conceptual resources for humanity as the search for extraterrestrial life – and most especially, extraterrestrial *intelligence* – continues. A long-standing concern, often invoked in science fiction, about any human encounter with intelligent extraterrestrial life is that such an encounter would lead to widespread panic and chaos. If the idea of extraterrestrial life contradicts deeply held religious beliefs, then proof of its existence could very well undermine social cohesion in unpredictable ways that could potentially threaten civilization, pulling the proverbial rug from under large swathes of humanity.

This is a significant point because it is not obvious that religion should be an ally in the search for intelligent life on other worlds. One of the classic science fiction tropes in stories and films about human encounters with extraterrestrial intelligence is the image of a human species unprepared for such encounters due precisely to fears rooted in religious prejudices. In the Asylum Films 2005 cinematic version of H. G. Wells' pioneering 1898 novel *The War of the Worlds*, for example, as well as in Jeff Wayne's 1978 musical adaptation of the same work, a Christian pastor describes the invading aliens as "demons." The idea of extraterrestrial life forms as demonic beings is present not only in science fiction, but in actual Christian discourse, such as in the work of author Guy Malone (1999).[1] In the year 1600, Giordano Bruno was sentenced to death by the Roman Inquisition and executed for, among other things, his belief in the existence of multiple inhabited worlds in the universe, along with his pantheism and his belief in reincarnation (Martinez, 2018).

To be sure, there have also been Christian thinkers who have entertained or even welcomed the idea of intelligent extraterrestrial life. Interestingly, as early as the thirteenth century, St. Thomas Aquinas (1225–1274) seriously entertained the possibility of intelligent extraterrestrial life in the cosmos. While he concluded, based on the scientific knowledge available at the time, that such life was improbable, he did not regard it as, in principle, impossible, or as violating any basic principles of Christianity (George, 2001).

The idea of extraterrestrial intelligence has become more theologically pressing in the modern era, during which the possibility and even the likelihood of the existence of such intelligence has come to be taken ever more seriously by intellectuals and has entered mainstream consciousness through the pervasiveness of science fiction in popular media. Most recently, the wildly successful search for exoplanets has made the imminent discovery of extraterrestrial life appear almost inevitable.

Probably the first prominent Roman Catholic theologian and scientist, Pierre Teilhard de Chardin (1881–1955), to engage with this concept also affirmed, like Thomas Aquinas before him, the view that intelligent extraterrestrial life could very well exist, still taking this possibility to be unlikely, but giving it greater credence due to the more advanced knowledge of the cosmos available from modern science:

As early as 1920 in his essay "Fall, Redemption and Geocentrism" Teilhard had taken notice of the probability of there being other galaxies in the universe, a likelihood which was confirmed by Hubble just a few years later, and he began to wonder what other possibilities of life they might harbor. Although…Teilhard appears at first to have confined himself to describing the evolution of life on our planet and to have expressed the opinion in his 1945 lecture on "Life and The Planets" that life-supporting planets might be exceedingly rare, still, he suggests that, even if this should be true, to discount the phenomenon on the basis of the comparative rarity of life in the universe may be a sign that we are looking through the wrong end of our telescope!

(Kropf, 2000)

The renowned Protestant theologian and science fiction and fantasy author C. S. Lewis also took the idea of intelligent extraterrestrial life quite seriously and also recognized that the possibility of such life raises significant questions for Christian theology (Earls, 2023). Would non-human intelligent beings stand in need of Christ's redemption? Would intelligent extraterrestrials have their own Christs (George, 2001)?[2] Is redemption as conceived in Christianity the only possibility available to God for saving living beings, or is such redemption unique to humanity?

All the ideas discussed here, even if entertained and taken quite seriously by respected Christian thinkers, have tended to fall outside the mainstream of Christian thought, at least before the modern period (though to some extent even today). The execution of Giordano Bruno is testimony to the fact that there have been periods of Christian history in which such ideas have been regarded not only as eccentric but also as dangerously heretical.

Extraterrestrial Life and the Dharma Traditions: Potential Implications

The kinds of ideas for which Bruno was put to death, though – ideas like the existence of multiple inhabited worlds in the universe, pantheism, and reincarnation – are mainstream in Hindu traditions and have been so for many centuries.

To be sure, the existence of extraterrestrial life does not necessarily imply the truth of pantheism or the reality of reincarnation.

There are, however, some interesting affinities among these ideas. The question of extraterrestrial life raises, among other issues, the question, "What is life?" In our search for other intelligences, are we going to confine our attention to life forms that are recognizably like ourselves? While there may be pragmatic reasons to limit our search in this way, at least in its preliminary stages (such as by focusing on planets with oxygen atmospheres), might such limitations cause us to miss forms of intelligent life that do not fit our preconceived mold of what life is? Is this not simply a further extension of our unfortunate human tendency toward ethnocentrism? This question calls to mind the *Star Trek* Original Series episode "The Devil in the Dark," in which a silicon-based life form called the Horta is discovered. The Horta, whose offspring have been mistaken by human miners for inanimate silicon nodules and indiscriminately destroyed by them, goes on a murder spree to stop this destruction. What is first thought to be a monster is found to be a protective mother, and silicon nodules that were not even recognized as life forms are found to be her eggs.

"The Devil in the Dark" suggests that the human propensity to fail to see potential life everywhere and to respect it accordingly could end up having a karmic rebound effect upon us, with our own destructive tendencies leading to our destruction. Such a sentiment would find ready agreement in the Jain tradition, which affirmed the existence of tiny life forms inhabiting air and water millennia before the invention of the microscope.

According to Jain teaching, there is literally life everywhere:

> Living reality, or *jiva*, is broadly defined as dynamism and suffuses what in pre-contemporary physics would be considered inert. Each jiva is said to contain consciousness, energy, and bliss. Earth, water, fire, and air bodies (which comprise material objects such as wood or umbrellas or drops of water or flickers of flame or gusts of wind) all contain jiva, or individual bodies of life force.
>
> *(Chapple, 2001)*

With the existence of life, according to Jain teaching, comes an obligation to protect that life from harm. This is the principle of *ahiṃsā*, or nonviolence in thought, word, and action which comprises the central ethical principle of the Jain tradition, and of the Hindu and Buddhist traditions as well.

Belief in reincarnation or rebirth (*punar-janma* in Sanskrit) is also central to the worldviews of the Hindus, Buddhists, and Jains. That which is reborn is called, as we have just seen, the *jiva*, or life force, in the Jain tradition, as well as in the Hindu tradition. Buddhist traditions tend to speak more in terms of a flow of discrete consciousness events, rather than of a singular "soul," for reasons that have to do with Buddhist soteriology (Collins, 1990).[3] The important point here, in relation to extraterrestrial life, is that there is nothing specifically *human* about the jiva. All living beings, by definition, possess – or rather, *are* – jivas. Bodies are merely the vehicles through which jivas experience the results of their actions – their karma – and engage in further actions, leading to further results, and so on. If it is the karma of a living being to be reborn in a human body, it will be reborn in a human body. If, however, its karma would be better fulfilled through rebirth in another form, it will be reborn in another form. In this sense, the Dharma traditions are not anthropocentric. There is nothing that is fundamentally human about the jiva. There are human bodies, cat bodies, dog bodies, ant bodies, cow bodies, and presumably extraterrestrial bodies of many types. A single jiva, however, may be born in any of these forms, or all of them and more, in the course of its journey. As the *Bhagavad Gita*, a major Hindu scriptural text, proclaims:

> Never was there a time when I did not exist, nor you...nor in the future shall any of us cease to be. As the embodied soul continually passes, in this body, from boyhood to youth, and then to old age, the soul similarly passes into another body at death. The self-realized soul is not bewildered by such a change...That which pervades the entire body is indestructible. No one is able to destroy the imperishable soul...For the soul there is never birth nor death. Nor, having once been, does he ever cease to be. He is unborn, eternal, ever-existing, undying and primeval. He is not slain when the body is slain...As a person puts on new garments, giving up old ones, similarly the soul accepts new material

bodies, giving up the old and useless ones…Knowing this, you should not grieve for the body.

(Prabhupada, 1972)[4]

The soul – the jiva – not being human, can take on any type of body that it needs in order to receive the results of its karma and continue on its journey to self-realization. To paraphrase religion scholar, Arvind Sharma, "We are not human beings having a spiritual experience; we are spiritual beings having a human experience" (2013, pp. 3–4).

Again, as it relates to extraterrestrial life, given the vast size of the cosmos as taught in the Dharma traditions, as well as the fact that the cosmos is also taught to contain numerous inhabited "worlds," or *lokas*, it becomes clear that, from the perspective of these traditions, the idea that a human life form could easily be reborn in an extraterrestrial body, and vice versa, not only is not surprising, but it is almost to be expected. This is indeed an answer to the common objection to the idea of rebirth that it does not account for the increase through time of the (human) population.

The eleventh- to twelfth-century Hindu philosopher, Ramanuja, teaches that the universe is the body of God, and that God is the soul of the universe. This is quite close to Bruno's pantheism: the idea that God is ultimately identical with all of existence (Roberts, 2017, p. 57). Ramanuja's – and the wider Hindu tradition's – understanding of the relationship of God and the cosmos is better characterized as a form of *panentheism*, the idea that God is present within the totality of, but is not exhausted by, the cosmos. Indeed, Ramanuja's idea harkens back to the very ancient image in the *Rig Veda*, the oldest of the Hindu scriptures (and one of the oldest known scriptures in the world) of the world being formed from the body of a divine being: the *Purusha*, or "Cosmic Person." In the *Rig Veda*'s *Purusha Sukta*, or "Hymn of the Cosmic Person," it is said that when the deities formed the cosmos from the body of the Cosmic Person, only one-fourth of the Cosmic Person became the visible cosmos: "All creatures are one-fourth of him, three-fourths eternal life in heaven. With three-fourths Purusha went up: one-fourth of him again was here" (Griffith, 2009).[5] This is also more consistent with panentheism than with pantheism, which is widely taken to refer to an exhaustive *identification* of the cosmos with God, and even as a position akin to atheism, inasmuch as it denies that there is a being – God – that exists apart from the cosmos. Panentheism affirms that the reality of God is vastly more than that portion which manifests in and as the cosmos. It is the idea, found in many of the world's religions, of a Supreme Being who is both transcendent and immanent, both outside of, and present within, the realm of time, space, and causation.

The relevance of panentheism to the question of extraterrestrial intelligence is its implication that the entire cosmos is pervaded by the divine consciousness. From this perspective, it would be more surprising if there were *not* intelligences scattered throughout the universe than if there were. In a sense, from this perspective, the cosmos itself could be said to constitute a living thing: the "body" of which the divine consciousness is the "soul," as in Ramanuja's teaching.

The concept on which both Thomas Aquinas and C. S. Lewis speculate in the Christian tradition, and that also contributed to the condemnation of Bruno – the idea of many divine incarnations – is a mainstream concept in Hindu traditions in the form of the doctrine of the *avatar* or divine descent. Analogous concepts are also present in the Buddhist and Jain traditions in the form, respectively, of the many Buddhas and Bodhisattvas who manifest themselves in the world in order to help other living beings on their path to liberation from suffering and rebirth, and the Tirthankaras, who also appear in the world in order to re-establish the Jain path to liberation once it has declined in our world.

Giordano Bruno, in fact, was skeptical of the need for a divine atonement for human sin (Acocella, 2008). But his view that God was present everywhere, in all beings, lent itself to an openness to multiple divine incarnations, given that we are all, in a sense, such incarnations. This is a Hindu view as well.

While the Hindu tradition does not tie the concept of the avatar to life on other worlds specifically (though it certainly does not deny the possibility that divine incarnations could occur wherever they are needed), it does affirm that God can take on a living form (which does not necessarily have to be a *human* form) in order to achieve some specific goal in the upholding of the good – or *dharma* – in the universe. God – the Supreme Being referred to variously as *Ishvara* ("the Lord") or *Bhagavan* ("the Blessed One") in Hindu texts – has many forms or aspects. Each form plays a particular role in the ongoing divine activity of creating, sustaining, and occasionally destroying and recreating, the world (meaning the cosmos, the totality of manifest being). These divine forms are simultaneously forms of a singular Supreme Being and individual deities in their own right. Hinduism is thus both monotheistic and polytheistic (and, as already mentioned, panentheistic). The deity, or form of divinity, tasked with preserving dharma – the good, or the order of the world – is Vishnu. It is Vishnu who manifests in the world, periodically descending from the celestial realm to the realm of incarnate beings. Indeed, the term for these manifestations – *avatar* – literally means *descent*. One of the most prominent avatars of Vishnu, Krishna, is a major character in one of the two main Hindu epic stories, the *Mahabharata* (and the chief protagonist of the other epic, the *Ramayana*, is the avatar prior to Krishna, Rama). A central moment in the *Mahabharata* occurs just before the eighteen-day battle between the heroes of the epic – the Pandava brothers – and their rivals who have unjustly dispossessed them – the Kauravas. Just as the battle is about to begin, the leader of the Pandava army, Arjuna, has a moment of doubt regarding whether it is right for him to fight in it. The Kauravas are the cousins of the Pandavas, and the members of the Kaurava army against whom Arjuna must fight include figures who are very dear to him, such as his teacher, Drona, and the patriarch of the entire clan, Bhishma. Arjuna therefore asks for advice from his best friend and charioteer, Krishna. The ensuing dialogue has become a sacred Hindu text in its own right: the *Bhagavad Gita*, or "Song of God [Bhagavan]," which is a summary of Hindu philosophy. In the fourth chapter of this text, Krishna explains to Arjuna the concept of the avatar: "Whenever order declines and chaos arises, I manifest myself. To destroy evil and restore the good I appear age after age" (Bhagavad Gita 4.7–4.8. Translation mine).

It is not much of a stretch to read into these verses the idea that whenever *and wherever* order declines and chaos arises, an avatar appears. In other words, avatars might appear on worlds and in realms other than the Earth.

The concept of avatars is more fully developed in another Hindu text which focuses primarily on the life story of Krishna: the *Bhagavata Purana*. In this and subsequent texts, a narrative develops of a series of avatars, many of which are not human. The avatars include a fish (who plays a role in the Hindu equivalent of the Great Flood narrative), a great tortoise, a giant boar, a creature that is part human and part lion, and a dwarf-like humanoid.[6]

The idea of spiritually exalted or enlightened, beings manifesting not only at many different times but also on various worlds, is made explicit in the tradition of Mahayana Buddhism. Texts in this tradition refer to "worlds as numerous as the grains of sand of the Ganges" (Tegchok, trans. 2017). Bodhisattvas, beings who are on the verge of becoming free from the cycle of rebirth – the shared ultimate goal of Hindu, Buddhist, and Jain traditions – are held in Mahayana Buddhism to have the aspiration of liberating all of the beings in the cosmos, including all of those dwelling on the many worlds which make it up.

Hindu, Buddhist, and Jain texts speak abundantly about intelligent non-human living beings which inhabit various worlds. *World* in this context typically means something like a plane of existence, a realm in some way spatially distinct from the Earth. The Sanskrit term translated as *world*, *loka*, can also be translated as *planet*, *universe*, or *plane of existence*. "It can also be interpreted as a mental state that one can experience" (Lin, 2022). In the Hindu, Buddhist, and Jain traditions, there are many variations on the theme of lokas, and a variety of specific models of how many lokas exist, what their characteristics are, and their relationship to the Earth. Broadly speaking, all three traditions speak of celestial lokas – realms that are conceived as superior to the earthly realm, based on such factors as the exceedingly long lifespans of the beings who inhabit them, the varied powers of these beings, and the sheer beauty of these realms. All three traditions also speak of hellish realms, rebirth in which is conceived as a result of very bad karma, a consequence of deeply harmful and immoral action. The Earth is generally conceived as being part of – but not as coextensive with or as exhaustive of – a middle realm, between the celestial and hellish lokas.

To this extent, the Hindu, Buddhist, and Jain concept of many worlds seems to bear far more of a resemblance to classical Christian models of heaven, hell, and purgatory than to modern models of the cosmos as containing countless planets orbiting a similarly countless number of stars. And to be sure, these cosmological models seem to serve a similar role in these traditions to the images of hell and paradise found in Christianity (as well as Islam): as an inducement to good behavior in the present and avoidance of immoral actions that could lead one to an unpleasant afterlife. What appears to differentiate the Hindu, Buddhist, and Jain models, however, is, at least, in part, the very massive scales of both time and space that are involved in the conceptualization of these models. In contrast with the *Divina Commedia* of Dante Alighieri, in which hell is conceived as existing beneath the surface of the Earth (and thus within a relatively confined space), the hells and heavens and even the middle realm conceived in Hindu, Buddhist, and Jain texts are vast. In Jain sources, for example, the width of the middle realm where the Earth exists is described as being one *rajlok* in diameter. A rajlok is defined as equivalent to the distance covered by a deity flying nonstop in a *vimana*, or flying ship, for six months at a speed of 2,057,152 yojanas (roughly 10,000,000 miles) per second (JAINA, no date). Ten million miles per second is of course more than fifty times the speed of light.

In science fiction terms, Jain sources are essentially claiming that deities travel at warp speed. If such speeds were possible, in six months, one would cover roughly one hundred fifty trillion miles, or about twenty-six light years.

Hindu texts also depict a cosmos that is vast in both spatial and temporal terms. As Carl Sagan famously noted:

> The Hindu religion is the only one of the world's great faiths dedicated to the idea that the Cosmos itself undergoes an immense, indeed an infinite, number of deaths and rebirths. It is the only religion in which the time scales correspond, no doubt by accident, to those of modern scientific cosmology. Its cycles run from our ordinary day and night to a day and night of Brahma, 8.64 billion years long, longer than the age of the Earth or the Sun and about half the time since the Big Bang. And there are much longer time scales still.
>
> *(1980, p. 258)*

A correction that needs to be made to Sagan's account is his claim that the Hindu religion is "the only religion" that affirms such vast timescales. This is actually true of the Dharma traditions as a whole, including Jainism and Buddhism. Sagan's claim, of course, that these timescales are in close accord with those of modern scientific cosmology is "no doubt by accident" is one which many Hindus would, of course, contest, believing that the ancient seers, or *rishis*, were able to perceive the true nature of reality while deep in meditation.

The idea that the cosmos is not only incredibly old but that it also contains many inhabited worlds is also commonplace in the writings of the Indic traditions. We have already seen the Buddhist reference to "worlds as numerous as the grains of sand of the Ganges." (Tegchok, trans. 2017). Hindu texts refer to numerous worlds, or *lokas*, inhabited by various deities and other intelligent non-human beings. It is also believed by some that these worlds can be visited astrally or etherically by means of various meditative practices (see for example Prabhupada, 1970). One Buddhist school of thought, the Yogachara system of philosophy, is held to have been revealed to its founder, a monk named Asaṅga, by the bodhisattva Maitreya, who was visiting our world from his usual plane of existence, a celestial realm called the Tushita (the "happy" or "contented") heaven (Williams, 1989, p. 80).

Whether it is by accident or through some special insight of the ancient seers, the Dharma traditions are well equipped with cosmological concepts that would seem to smooth the way for the eventual discovery of extraterrestrial intelligence.

Thus, the notion that human religions would be shaken to their foundations by this discovery would seem to break down when we bring the Dharma traditions into our reflections. Such a discovery might even be seen by some practitioners of these traditions as validations of long-held beliefs.

Ancient Aliens? Not Necessarily

The question of whether the seers of the Dharma traditions truly perceived such things as the age and size of the cosmos and the presence of microscopic organisms in the air and on surfaces (the view of many adherents of these traditions) or whether they "got lucky," appearing to know these things, in Carl Sagan's words, "by accident" raises other questions as well.

With regard to the existence of extraterrestrial life, it raises the question of whether, if the Dharma traditions have been right about so many other things, they are also right about the existence of non-human intelligences, such as the *gandharvas*, *rakshasas*, *asuras*, and *devas* with which Hindu, Buddhist, and Jain texts are populated.[7]

It raises another possibility as well. If one is skeptical of whether sages in ancient India were able to peer so deeply into the nature of existence as to give a broadly accurate picture of reality, as it is now understood by contemporary science, but if one is also skeptical that ancient Indian accounts could also be so accurate purely as a matter of accident, the question arises as to whether ancient India was visited by extraterrestrial life forms who revealed this information to human beings who then preserved it in such texts as the Vedas (Kuepferle, 1986).[8]

This is of course the ancient astronaut hypothesis, which has been put forward by such authors as Erich von Däniken and popularized in the History Channel series *Ancient Aliens* (1971).[9]

I would like to suggest that a good stance to take toward this hypothesis is one of what I call "open-minded skepticism." Given, as many have argued in the past, the vast size of the Universe and what is increasingly being observed as the fairly commonplace nature of planets like the Earth, to rule out as impossible the idea that intelligent extraterrestrials might have visited our planet in the past, seems excessively dogmatic. The vast distances involved in interstellar travel, of course, militate against such a possibility, but they do not rule it out altogether. Given enough time, an artificial craft could navigate the distances between stars. The Universe, though, has existed for billions of years, and only a few thousand years would be needed for a craft to travel between two relatively close star systems at speeds below that of light. Such a craft might have the form of an artificial intelligence capable of surviving for such a lengthy period or be crewed by such intelligences.

It is also possible that an extraterrestrial species might have developed such great longevity as to be able to make the trip in person. Cryogenics might also be deployed to make such a trip feasible, and relativistic time dilation would also make the trip itself much shorter for the crew of a spacecraft traveling near the speed of light. All of these possibilities have been envisioned in science fiction, as has the idea of warp travel or wormholes that could allow a sufficiently advanced species to circumvent the light-speed problem (without technically violating the dictum that one cannot travel faster than light).

The idea that extraterrestrial visitors might have come to the Earth in ancient times is thus not at all impossible.

Is it, however, likely? And more to the point, is it *necessary* to postulate extraterrestrial visitors in order to account for the kind of evidence that is often proposed in favor of this concept, particularly with regard to India?

In regard to the last question, I remain skeptical of this hypothesis because other explanations are available to explain the truly interesting evidence that exists in ancient Indian sources.

First, the fact that ancient Indian sources talk about craft that can travel fifty times faster than the speed of light does not mean that such craft, in fact, exist. Contemporary science fiction also talks about such craft, along with matter transporters that can "beam" a person from a spacecraft in orbit to the surface of a planet (and back) and battle stations with enough firepower to destroy an entire planet. None of these things exist. We can envision them, however, and write fairly detailed stories about them because we are intelligent and imaginative creatures. We can even be inspired by such stories to build actual devices like those about which we have read or which we have seen in films or on television. The human brain has not changed so dramatically over the course of the last few thousand years – the period of the duration of human civilization – as to suggest that ancient peoples were incapable of also imagining, sometimes in great detail, ships that flew through the air and weapons with devastating power, as well as non-human intelligences. The fact that texts from the Dharma traditions describe aircraft, and even spacecraft, and that there is a Sanskrit word – *vimana* – to denote such craft is certainly extremely interesting.[10] It tells us that the ancient Indians were at least as intelligent and imaginative as people today. It does not tell us, however, that such devices actually existed, or that the authors of the ancient Indian texts must have seen such devices in order to be able to write about them in as much detail as they do. To suggest – and this strikes me as the chief weakness in the ancient astronaut hypothesis – that ancient people had to have seen spaceships in order to describe them is to sell ancient people short. If we can imagine hyperdrives and Death Stars and transporters, which we have never seen, why couldn't they?

Similarly, with regard to the actual technological achievements of ancient peoples (which also give us an idea of how intelligent human beings have long been), the ancient astronaut hypothesis suffers from the assumption that ancient human beings lacked the capacity to create structures like the pyramids of Egypt and Mesoamerica. This assumption smacks of colonialism. Indeed, science fiction legend Isaac Asimov attributes the ancient astronaut theory to the same impulse that led ancient conquerors to attribute the technological achievements of the peoples they had conquered not to those people, but to giants:

> Being innocent of the advanced technology of the civilization they have conquered, they cannot imagine how those structures were made. They themselves could not have done it, and it would therefore be ridiculous to suppose that the inferior people they had conquered could have done it. The logical assumption was that a race of giants did it...And it's not only naïve ancients who believed this. Some people today, surveying the pyramids of Egypt and convinced that the ancient Egyptians could not have built them, fantasize their own version of giants and demigods as having built them. They naively suggest that astronauts from other worlds did the job. (Why astronauts, with technologies capable of interstellar flight, should have constructed huge piles of stone rather than having built something of steel and concrete beats me.)
>
> *(1996, p. 182)*

The ancient astronaut hypothesis does indeed come dangerously close to the tendency that Asimov describes. It robs people like the ancient Indians (and their descendants) of agency in affirming that they could not possibly have done the amazing things which they actually did (and that they could not possibly have imagined the amazing things which they actually imagined).

Extraterrestrial Religion? Cosmic Pluralism and the Ideal of Acceptance[11]

According to *pluralism*, a stance increasingly embraced by Western theologians and philosophers of religion, there is not just one true worldview or way of life to which all must adhere. The idea is that the world's religions and philosophies represent many valid and effective paths that human beings might follow.

Pluralism and the related stance of *inclusivism*, which affirms one finally true worldview but is open to the truth in many perspectives, are both widely present in the Dharma traditions. Indeed, the pluralism of Hindu thinkers such as Sri Ramakrishna, Swami Vivekananda, and Mohandas K. Gandhi has been among the elements that have influenced Western philosophers to take this idea seriously.

While the idea of intelligent life on many worlds does not directly imply the truth of pluralism, it would seem to be consistent with such a view. Given the diverse religions human beings have developed, one wonders how much more diverse the religions of intelligent non-human species might be. If the pluralistic assumption that no culture has a monopoly on truth is valid, then a *cosmic pluralism* would seem to be a logical approach to take toward the potential religious diversity of the universe.

This, of course, begs several questions. Should we assume an intelligent extraterrestrial species would have religion at all? In many of the worlds envisioned by science fiction, religion is viewed negatively, as a peculiarly human malady, and one an advanced extraterrestrial civilization would have long shed by the time it contacted humanity. In this view, we humans, ourselves, will also outgrow our need for religion if we survive our current crises and continue to evolve a more rational sensibility.

Indeed, some might argue that leaving religion behind is a precondition for human survival. Many science fiction authors have believed as much. Gene Roddenberry conceived of the future depicted in *Star Trek* as a wholly secular one (though subsequent contributors to *Star Trek* lore have deviated from this vision, especially *Star Trek: Deep Space Nine*, in which religion plays a central role).

On the other hand, even if extraterrestrial religions are radically different from human religions, perhaps so different as to stretch the very meaning of the word *religion*, certain universal questions could be salient even for non-human life forms. Death and loss, for example, form a powerful motivation to develop beliefs about an afterlife, and practices designed to create meaning in the face of mortality. And human traditions alone are so diverse as to stretch and challenge conventional meanings of the word *religion*. Is Confucianism (or Ruism) a religion? Much depends on how the word *religion* is defined. If we are willing to think broadly of what we might call spiritual aspirations, there seems to be no reason to assume these will be absent in intelligent extraterrestrial beings, even if the quest to discern them might prove to be among our greatest intellectual challenges, should we have the opportunity to pursue it.

Though he does not address the idea of extraterrestrial life, the modern Hindu teacher, Swami Vivekananda (1853–1902) does speak about another science fiction theme when he raises the question of the religions human beings might practice in the future. He affirms pluralistic convictions, based on Hindu philosophy, which can provide some guidance in thinking about possible extraterrestrial religions. He first affirms a strongly pluralistic attitude toward the past and present religions of humanity:

> accept all religions that were in the past, and worship with them all; I worship God with every one of them, in whatever form they worship Him. I shall go to the mosque…I shall enter the Christian's church and kneel before the crucifix; I shall enter the Buddhistic temple, where I shall take refuge in Buddha and in his Law. I shall go into the forest and sit down in meditation with the Hindu, who is trying to see the Light which enlightens the heart of every one.
>
> *(Vivekananda, 1979, p. 374)*

He then affirms the same attitude toward religions that do not yet exist:

> Not only shall I do all these, but I shall keep my heart open for all that may come in the future. Is God's book finished? Or is it still a continuous revelation going on? It is a marvellous book–these spiritual revelations of the world. The Bible, the Vedas, the Koran, and all other sacred books are but so many pages, and an infinite number of pages remain yet to be unfolded. I would leave it open for all of them. We stand in the present, but open ourselves to the infinite future. We take in all that has been in the past, enjoy the light of the present, and open every window of the heart for all that will come in the future. Salutations to all the prophets of the past, to all the great ones of the present, and to all that are to come in the future!
>
> *(Ibid.)*

It is completely in keeping with the Hindu tradition's affirmation of the reality of many worlds and forms of life to extend the pluralism that Vivekananda affirms here toward the religions of the future to the possible religions of intelligent extraterrestrial civilizations. This extension could occur based on the principles Vivekananda affirms: that "We take in all that has been in the past, enjoy the light of the present, but open every window of the heart for all that will come in the future." From a human perspective, any extraterrestrial religion would be, in our current time, a religion of the future, as we have yet to contact intelligent extraterrestrial life.[12]

My claim is that the search for extraterrestrial life needs to be accompanied by the very openness to diverse views and possibilities that Vivekananda commends. If there are many extraterrestrial civilizations, there would seem to be every reason to expect that the religious diversity of the universe would also be correspondingly vast. It may well be, as some science fiction predicts, that there are extraterrestrial species that have left religion behind, or that have never had it. But it may well also be true that there are, as other science fiction works have predicted, as much variety of religious beliefs among extraterrestrial civilizations as exists on our single small planet today. We would do well to be open and accepting toward extraterrestrial perspectives, no less so than human ones.

Conclusion

The possibility of discovering and, perhaps one day, even encountering intelligent extraterrestrial life is a most exciting one, to say the least. The search for such life is an extension not only of our natural human curiosity but also of our desire to communicate with and understand other beings, and ideally learn from them as well.

The religious quest, too, is, among other things, a search to expand our awareness, and to contact something greater than ourselves, whether it be conceived as a deep potential within ourselves or as the intelligence from which the cosmos itself has emerged.

Religious beliefs, like any beliefs, can imprison us. They can be held dogmatically and preclude other possibilities, often to the detriment of human flourishing. But they can also inspire us and cause us to seek out new vistas of knowledge and expression.

The Dharma traditions suggest several ways in which these two searches might enable one another. Religious belief need not be a barrier to accepting the possibility of extraterrestrial intelligence. It can be an affirmation of that possibility and an encouragement as we pursue this quest in earnest.

Notes

1 See, for example, Malone's book, *Come Sail Away: UFO Phenomenon and the Bible* (Seekyel Com Online Publishing, 1999), and his video, *Are Aliens Demons? Evidences for a Spiritual View* (2013).
2 Interestingly, Thomas Aquinas, too, examines the possibility of multiple divine incarnations. He concludes that God could very well choose to incarnate in multiple forms, though he also asserts that there is no basis in divine revelation for believing that this has in fact happened. See George (2001).
3 For a more in-depth discussion of this topic, see Steven Collins, *Selfless Persons: Imagery and Thought in Theravada Buddhism* (Cambridge: Cambridge University Press, 1990).
4 *Bhagavad Gita* 2:11b–13, 17, 20, 22, 25b.
5 *Rig Veda* 10.90.3b–4a.
6 Some modern Hindu teachers, most notably Sri Aurobindo (Aurobindo Ghose, 1872–1950), have suggested that the non-human avatars and the order in which they are said to appear are symbolic of the evolution of life, passing through ocean-dwelling forms, then to reptilian forms, then to mammalian forms, and then to humanoid and human forms.
7 These are just a few of the many species of intelligent non-human beings described in texts of the Dharma traditions.
8 It should be noted that there are elements of the cosmos described in the Dharma traditions that are not so easily reconciled with a contemporary scientific cosmology. The Jain conception of the world disc, for example, despite the fact that this disc is described as, in today's terms, twenty-six light years in diameter, has been taken by at least some Jains to be an affirmation of the notorious "Flat Earth Theory." This has sometimes led to crises of faith, such as that undergone by former Jain monk Kumar Jayakirti, who left monastic life and become a householder upon hearing the launch of Sputnik and having his faith in Jain cosmology shaken by the demonstrable fact that the Earth is round. See Paul L. Kuepferle, *The Frontiers of Peace: Jainism in India* (BBC TV, 1986).
9 See Erich von Däniken's *Chariots of the Gods?* This text remains the classic statement of the ancient astronaut idea.
10 The most fascinating thing about the word *vimana* is that it is not based on an attempt to describe a flying machine based on an existing ancient technology. It does not, in other words, break down to mean "flying chariot" or "flying house." It just means flying craft, and nothing else.
11 I have previously explored ideas similar to the ones in this section in "Worlds as Numerous as the Grains of Sand of the Ganges," which is Chapter 11 in Andrew M. Davis and Roland Faber, eds., *Astrophilosphy, Exotheology, and Cosmic Religion: Extraterrestrial Life in a Process Universe.* Lanham: Lexington, 2024.
12 Again, at least beyond dispute, respectfully setting aside the many interesting claims that this has already occurred.

References

Acocella, Joan (August 18, 2008) "The Forbidden World," *The New Yorker*, Available at: www.newyorker.com/magazine/2008/08/25/the-forbidden-world Accessed February 1, 2024.

Asimov, Isaac (1996) *Magic: The Final Fantasy Collection*, New York: HarperCollins.

Chapple, Christopher Key. (2001) 'The Living Cosmos of Jainism: A Traditional Science Grounded in Environmental Ethics', *Daedalus*, Available at: www.amacad.org/publication/living-cosmos-jain sim-traditional-science-grounded-environmental-ethics Accessed February 1, 2024.

Collins, Steven (1990) *Selfless Persons: Imagery and Thought in Theravada Buddhism*, Cambridge: Cambridge University Press.

Davis, Andrew M. and Faber, Roland, eds. (2024) *Astrophilosophy, Exotheology, and Cosmic Religion: Extraterrestrial Life in a Process Universe*, Lanham: Lexington.

Earls, Aaron (June 16, 2023) 'C.S. Lewis Warned Us about Close Encounters of the Evangelical Kind', *Christianity Today*, Available at: www.christianitytoday.com/ct/2023/june-web-only/cs-lewis-warned-us-about-close-encounters-of-evangelical-ki.html Accessed January 18, 2024.

George, Marie I. (2001) 'Aquinas on Intelligent Extra-Terrestrial Life', *The Thomist*, Vol. 65, No. 2, pp. 239–258.

Griffith, Ralph T.H., trans. (2009) *Rig Veda*, Santa Cruz: Evinity Publishing (First published in 1896).

JAINA (no date) 'Jain Reality/Existence' JAINA: Federation of Jain Associations in North America, Available at: www.jaina.org/page/RealityExistence Accessed January 18, 2024.

Kropf, Richard W. (2000) 'Teilhard de Chardin's Vision of Ultimate Reality and Meaning in the Light of Contemporary Cosmology', *Ultimate Reality and* Meaning, Vol. 23, No. 3, pp. 238–259.

Kuepferle, Paul L. (1986) *The Frontiers of Peace: Jainism in India* (BBC TV).

Lin, Qian (2022) 'On the Early Buddhist Attitude Toward Metaphysics', *Journal of Indian Philosophy*, Vol. 50, No. 1, pp. 143–162.

Malone, Guy (1999) *Come Sail Away: UFO Phenomenon and the Bible*, Seekyel Com Online Publishing.

Martínez, Albert A. (March 19, 2018) 'Was Giordano Bruno Burned at the Stake for Believing in Exoplanets? Most Historians Say No, but New Evidence Suggests Otherwise', *Scientific American*, Available at: https://blogs.scientificamerican.com/observations/was-giordano-bruno-burned-at-the-stake-for-believing-in-exoplanets/ Accessed January 18, 2024.

Prabhupada, A. C. Bhaktivedanta Swami (1970) *Easy Journey to Other Planets*, Mumbai: Bhaktivedanta Book Trust.

Prabhupada, A. C. Bhaktivedanta Swami (1972) *Bhagavad Gita: As It Is*, Los Angeles: Bhaktivedanta Book Trust.

Sagan, Carl (1980) *Cosmos*, New York: Random House.

Sharma, Arvind (2013) *Gandhi: A Spiritual Biography*, New Haven and London: Yale University Press.

Tegchok, Khensur Jampa, trans., and Thubten Chodron, ed. (2017) *Practical Ethics and Profound Emptiness: A Commentary on Nagarjuna's Precious Garland*, New York: Simon and Schuster.

Vivekananda, Swami (1979) *Complete Works*, Mayavati: Advaita Ashrama.

von Däniken, Erich (1971) *Chariots of the Gods?* New York: Bantam Books.

Voss Roberts, Michelle (2017) *Body Parts: A Theological Anthropology*, Minneapolis: Fortress Press.

Williams, Paul (1989) *Mahāyāna Buddhism: The Doctrinal Foundations*. London: Routledge.

9

ABRAHAMIC PERSPECTIVES ON EXOPHILOSOPHY

The Jewish, Christian, and Islamic Traditions on the Possibility of Extraterrestrial Life

Richard Playford

Introduction

The Deist thinker Thomas Paine once wrote that:

> To believe that God created a plurality of worlds, at least as numerous as what we call stars, renders the Christian system of faith at once little and ridiculous; and scatters it in the mind like feathers in the air. The two beliefs cannot be held together in the same mind; and he who thinks that he believes both, has thought but little of either.
>
> *(1794, p. 40)*

Clearly for Paine the existence of alien life is incompatible with Christian belief. More recently, the physicist Paul Davies has expanded the predicted casualties amongst the world's religions should alien life be discovered. He writes:

> Undoubtedly the most immediate impact of an alien message would be to shake up the world's faiths. The discovery of any sign that we are not alone in the universe could prove deeply problematic for the main organized religions ... In fact, I would go so far as to say that the discovery of aliens would deal a severe blow not only to Christianity, but to all mainstream religions. I am not saying that what we may loosely call the spiritual dimension of human life would be eclipsed or belief in some sort of wider meaning or purpose in the universe negated. Buddhists would doubtless continue to seek the path of enlightenment through inner reflection, even when armed with the knowledge of intelligent life beyond Earth. What is clear, however, is that any theology with an insistence on human uniqueness would be doomed.
>
> *(2010, pp. 188 and 192–193)*

Of course, all three Abrahamic religions affirm the uniqueness of human beings: Judaism and Christianity in texts such as Genesis 1:27, which teaches that God created mankind 'in his own image', and Islam in texts such as Qur'an 2:30, which teaches that humans are

DOI: 10.4324/9781003440130-14

God's viceregents on Earth, and Qur'an 17:70, which teaches that God has conferred on human beings a special dignity. As a result, all three religions appear to be in trouble, but is this really the case?

At the risk of spoiling the ending, the purpose of this chapter is to argue that this is not the case. It will do this in three stages. In the first section of this chapter, I will begin by outlining some of the issues raised by alien life. Unfortunately, due to limits on space, I will not be able to respond to all of them in exhaustive detail, but I will be able to gesture towards the sorts of strategies followers of the Abrahamic faiths can adopt when responding to them. In the second section, I will turn to Christianity in particular and will focus on the issue of 'narrative conflict'. This section will have more a theological focus and will show how the Christian 'narrative' could respond to the discovery of alien life. The third and final section will be more philosophical in nature and will argue that it is not at all unlikely, given the Abrahamic conception of God, that he has created life elsewhere in the Universe.

A Square Peg in a Round Hole: Alien Life and Abrahamic Theology

Whilst Davies, as we've seen, identifies the doctrine of human uniqueness as incompatible with the existence of intelligent aliens, the 'problems' posed by alien life are much broader. (I use 'problems' in quotation marks because, as we shall see, they are not really problems.) All three Abrahamic faiths have a series of doctrines that attempt to describe God's relationship to the world in general and human beings in particular. At the same time, none of the holy texts of these religions explicitly discusses alien life,[1] nor do they describe how intelligent aliens would fit into this schema. As a consequence, should alien life, particularly of an intelligent sort, exist, then it will need to be accommodated within the theology of these faith traditions, but, as we've seen, we don't have an obviously prepreparared way to do this. Thus, it is not so much the case that we are trying to fit the square peg of alien life into a round hole within the theology of these traditions, rather we don't even know if there are any holes (of any shape!) available for them to begin with.

The good news is that theologians from all three traditions have not been idle, and through careful study, numerous ways of incorporating alien life into the theology of these traditions have been identified. To provide a complete overview of the various issues goes beyond the space available. This is particularly the case when we consider the fact that we can be more or less fine-grained when considering how aliens might fit into Jewish, Christian, and Islamic theology. If, for example, we simply asked whether alien life in general would undermine the Christian doctrine that God created everything *ex nihilo* (out of nothing), then the answer is obviously no. A Christian can simply affirm that if alien life exists, then God also created it *ex nihilo*. On the other hand, if we asked whether an alien could become a Roman Catholic priest, well, now we have a more difficult question on our hands! How do we understand the priesthood? What are these aliens like? Presumably they would have to be intelligent, but would they have to be male and celibate? Well, what if they reproduced differently to us? What if they had three sexes? Could only one of them become priests? If so, which one? Let's suppose we decided that to keep it simple we simply said they couldn't be priests. Well, could they at least be baptized? The Pope's astronomer has said that he would baptize an alien (Jha, 2010), but what if the alien's biochemistry is such that baptismal water would kill it? And so on. Perhaps some of these examples are a bit silly, although at the moment we have no idea what forms alien life might take so, perhaps not, but hopefully my point is clear enough.

Despite this, a number of key issues do seem to have been consistently identified, and explored, by theologians. For Judaism, these include questions about the uniqueness of man, the uniqueness of the creator, and the relationship between man and God (Lamm, 1965, p. 21), and whether an alien could ever become Jewish (Ioannides, 2019). For Christianity, Losch and Krebs (2015) highlight that issues might be raised for the doctrines of creation, revelation, and redemption (which includes issues of both incarnation and salvation within it). Peters (2011) also lists creation and redemption as potential issues, whilst also adding (in a similar vein to Lamm's discussion of the relationship between man and God) that alien life might challenge human uniqueness. To this list McIntosh and McNabb (2021) add that there may also be issues of narrative conflict. We will return to this shortly. Unsurprisingly, given all three faiths share an Abrahamic foundation, similar potential issues are identified for Islam. These include personhood and the uniqueness of humanity (Jalajel, 2024; Abdullah, 2024; Karamali, 2024), issues of eschatology (Montasseri, 2024), and I have recently argued that there may be issues for revelation, scripture, and Muslim practice (Playford, 2024; Playford, Bullivant, and Siefert, 2024).[2] The good news is that all of the sources just referenced for all three traditions also propose solutions to the problems they identify, and this should give us confidence that all three religions have the intellectual resources needed to respond to these issues. Sadly, there is insufficient space to dig into all of these issues here. However, for an overview of these issues from the perspectives of all three religions, see Playford, Bullivant, and Siefert (2024); for a really detailed dive into these issues from a Christian perspective, see Davison (2023); and for a really detailed dive into these issues from a Muslim perspective, see Malik and Determann (eds.) (2024).

We can clearly see that contemporary theologians are thinking about these issues carefully. However, it is worth noting that this is hardly a new phenomenon and theologians throughout history have wrestled either with these exact same questions or with ones very much like them. To give a few examples, the ninth century Frankish monk Ratramnus of Corbie (800–868 AD) wrote, at quite some length, about the *cynocephali* or dog-headed men. The text was written in response to a query by a contemporary of his, named Rimbert (c. 830–888 AD), a priest who was engaged in regular missionary activity in what is now Denmark and Scandinavia where the *cynocephali* were rumoured to live. Rimbert appears to have been uncertain as to whether, should he encounter the *cynocephali*, they would be appropriate targets for evangelization. The details of the argument do not matter for our purposes, but ultimately Ratramnus concluded that they were appropriate targets for evangelization (Bruce, 2006; Matter, 2006). Whilst the idea of dog-headed men may strike us as absurd, it is worth remembering that at that time much of the world was unexplored (just as much of the galaxy remains unexplored today), and that there is no necessary a priori reason why we should be the only rational species on this planet (or indeed in the galaxy). We should, therefore, not dismiss their belief in (at least the possibility of) the *cynocephali* as a sign of superstition, ignorance, or stupidity. For our purposes, the important thing to note is that whilst Ratramnus was not considering the implications of *extraterrestrial non-human intelligences*, he *was* considering the implications of *terrestrial non-human intelligences*,[3] and in this respect, his concerns were similar to those of contemporary (astro)theologians.

Jumping forward a few hundred years, we find two theologians both considering the exact topic of this chapter. The fifteenth-century philosopher, theologian, and cardinal Nicholas of Cusa (1401–1461 AD) directly considered the possibility of extraterrestrial life. He concluded that it likely existed and that it had a part in God's plan for creation. In his book *On Learned ignorance*, he wrote:

> We surmise, that in the solar region there are inhabitants which are more solar, brilliant, illustrious, and intellectual – being even more spiritlike than [those] on the moon, where [the inhabitants] are more moonlike than [those] on the earth, [where they are] more material and more solidified ... We believe this on the basis of the fiery influence of the sun and on the basis of the watery and aerial influence of the moon and the weighty material influence of the earth. In like manner, we surmise that none of the other regions of the stars are empty of inhabitants – as if there were as many particular mondial parts of the one universe as there are stars, of which there is no number.
>
> *(Quoted in Foltz, 2019, p. 360)*

At around the same time, the French theologian William Vorilong (ca. 1390–1463 AD) also considered the possibility of alien life. After putting forward some arguments for why we should believe in a plurality of worlds (although whether these are to be interpreted as physical worlds is less clear), he writes:

> If it be inquired whether men exist on that world, and whether they have sinned as Adam sinned, I answer no, for they would not exist in sin and did not spring from Adam. But it is shown that they would exist from the virtue of God, transported into that world, as Enoch and Elias [Helyas] in the earthly paradise. As to the question whether Christ by dying on this earth could redeem the inhabitants of another world, I answer that he is able to do this even if the worlds were infinite, but it would not be fitting for Him to go into another world that he must die again.
>
> *(In McColley and Miller, 1937, p. 388)*

Put another way, Vorilong believes that 'men' on other worlds are likely sinless, and thus in no need of an incarnated redeemer (i.e., Christ), but that if they did need a saviour, Christ's death on Earth would be sufficient to save them without the need for multiple incarnations. Once again, however, the details do not matter. The point for our purposes is that theologians have been wrestling with these questions for a long time.

Finally, before closing this section, I want to point out a very simple 'one-size-fits-all' strategy that (astro)theologians can adopt when considering the possibility of extraterrestrial life. Often the instinct when considering extraterrestrials is to try to fit them into the existing theology of the respective traditions. We see, for example, that Vorilong assumes that if aliens are sinful, then they need to partake in Christ's death and resurrection in order to be saved. He then considers how this might work. This is a perfectly reasonable strategy. One can simply go through various key doctrines of the religion under consideration and consider the different ways extraterrestrials might fit into them. However, there is an alternative strategy. Rather than trying to fit *aliens into the religion*, we could instead fit the *earthly religion into a universal cosmic religion.* The question now becomes: what might this strategy look like?

One excellent illustration of this strategy can be found in Ray Bradbury's short story 'The Fire Balloons' (2008). In it a group of Episcopalian priests want to evangelize a group of Martians. The Martians take the form of spheres of blue fire. As the priests travel towards the Martians, they discuss the nature of religion and sin,[4] as well as how best to spread the gospel to such an alien species (for example, they decide to represent Christ as an orb so as to better help the Martians understand the importance of the incarnation). They arrive, set up an altar, and then pray whilst waiting for the Martians to arrive.

Sure enough, eventually the Martians arrive. However, to the surprise of the priests, rather than enquiring about the meaning of the altar, orb, intentions of the priests, and so on, the Martians simply thank the priests for coming whilst also declaring that it was all unnecessary and that they have wasted their time. We then learn that the Martians are the 'Old Ones'. In the distant past they had bodies similar to ours, but through some unspecified means, they found a way to cast them off, taking on their current form which is free 'of ills and melancholies, of deaths and transfigurations, of ill humours and senilities ... neither prideful nor arrogant, neither rich nor poor, passionate nor cold ... we shall never die, nor do harm'. Crucially, however, they 'live in God's grace' and have 'left sin behind' (Bradbury, 2008, pp. 169–170). As the Martians begin to leave, one of the priests asks whether he can return to learn more from them. The Martians say that he can and then leave, after which the priests also return home to reflect upon their experience.

For our purposes what is interesting about the story is that the priests go into it wanting to incorporate these extraterrestrials into earthly Christianity. On the contrary, however, the Martians invite the priests, as Christians, into a greater cosmic religion. The Martians seem to affirm at least certain aspects of Christianity, and display no desire to 'correct' the priests' theology, whilst also affirming that they too are part of God's creation and that their salvation and relationship with God are provided for, albeit in a very different manner than that of terrestrial human beings.

This strategy is available to all three Abrahamic faiths. The scriptures of all three traditions, for example, are aimed at human beings, and as a result they describe human events and concern human conduct and belief here on Earth. As already stated, they simply don't discuss alien life (nor would we expect them to). This gives adherents of these faiths an enormous amount of flexibility when considering alien life. There is nothing to stop God from providing for different species in different ways, and thus having very different religions. Further, if God has done this, the limits on how he might have done that are limited only by God's imagination! The possibilities are endless.

Two main concerns may come to mind at this stage. The first is that such a strategy might imply a version of religious pluralism, such as the one advocated by John Hick (1922–2012). Religious pluralism is the view that each of the major world religions is in some sense of the word 'true', because all of them are simply different ways of understanding a core 'transcendent reality' (Hick, 2004, p. 2). Religious pluralists, of course, may welcome such an implication, but many thinkers have significant concerns about such a view (for example, see D'Costa, 2009, pp. 9–19), and this may cause them to be wary of adopting such a strategy.

However, such a concern is unwarranted, since adopting this strategy in no way obliges us to adopt (at least a strong version of) religious pluralism. A quote from Bradbury, again, helps to illustrate why:

> This Truth here [on Mars] is as true as Earth's Truth, and they lie side by side. And we'll go on to other worlds, adding the sum of the parts of the Truth until one day the whole Total will stand before us like the light of a new day.
>
> *(2008, p. 172)*

Note here that the 'Truth' on Earth is singular. This allows us to claim that not all earthly religions are equally true, and that there are singular answers to religious questions. There is nothing about such a view, for example, that would oblige us to hold simultaneously that Jesus *is* the Second Person of the Trinity (as per Christianity) and also that he *isn't* the Second Person of the Trinity and was just a prophet (as per Islam) or just a regular human being (as per Judaism). (When laid out like this, it becomes clear why many thinkers are suspicious of religious pluralism!) Returning to the analogy of a puzzle, just because there is more than one piece to a puzzle does not mean that more than one piece can be inserted into any particular slot. Thus, one way to think about this would be to distinguish between, say, earthly Judaism (or Christianity or Islam) and Cosmic Judaism (or Christianity or Islam). Only earthly Judaism (for example) will fit into the Cosmic puzzle and the other religions, at least as much as they disagree with Judaism, are mistaken. With this distinction in mind, we simply don't need to fit aliens into earthly Islam (for example), because we can simply trust that they will fit into Cosmic Islam, whatever that looks like.

This also hints at the next objection one might make of this strategy (of how to respond to the possibility of alien life). One might object that this strategy explains nothing, since we haven't explained what this Cosmic religion looks like nor how extraterrestrial life fits into it (even if we do understand how we fit into it with our earthly religion). The problem with this objection is that it assumes that we must know everything before we can know something, which is obviously absurd. A Christian theologian, for example, is perfectly justified in saying we don't know how aliens fit into God's plan for the Universe, whilst also claiming that we *do* know how *we* fit into God's plan for the Universe. This is no different than any other field of human enquiry. There are many questions in science and history to which we do not know the answer, but this doesn't stop us from knowing the answer to some questions! It is unfair to demand more from the theologian than we would from an expert in any other field of human enquiry (particularly before we have even discovered alien life)!

Narrative Conflict and Subverted Expectations

In the previous section, we saw that theologians from all three Abrahamic faiths have thought carefully about the implications of alien life for their respective traditions. An overview of some of the key issues that have been identified was provided, and we then saw a general strategy which will allow us to respond to all of those concerns. In this section, I want to return to one of the issues listed earlier, namely, that of narrative conflict, whilst also considering one strategy Christians, and to a lesser extent Jews (and possibly Muslims), might adopt when trying to respond to it.

As McIntosh and McNabb (2021) point out, Christianity provides us with a narrative with which to understand the world. They write 'it is natural to think of Christianity as, in fact, a story of sorts. It is a grand narrative about life, the universe, and everything' (2021, p. 13). Different Christians will articulate this narrative in different ways, but certainly God's love for mankind (John 3:16) and our response to that love are key components. The Christian story, then, could be seen as a love story between God and humanity.

However, the discovery of alien life, depending upon how it happens, could disrupt that narrative. McIntosh and McNabb point out that how much tension is generated will depend upon what forms our interactions with alien life take. They suggest that the following five scenarios generate increasing tension:

S1. ETI is so remote or undetectable that any interaction is (nomologically) impossible.
S2. ETI is so remote as to be physically inaccessible, but communication is possible.
S3. Physical interaction with ETI is possible; ETI is peaceable.
S4. Physical interaction with ETI is possible; ETI is hostile but not an existential threat.
S5. Physical interaction with ETI is possible; ETI is hostile and an existential threat.

(2021, p. 14)

They point out that S1 and S2 generate little narrative conflict, but S4 and S5 seem to generate a lot. They write:

And so here, finally, we think we encounter a source of potential conflict between the existence of ETI and Christian belief. It is conceivable to us that ETI disrupts the Christian story, so to speak. It would be quite jarring, for example, if UFOs started appearing in the sky at the denouement of the cinematic masterpiece *The Shawshank Redemption*, although there is nothing in the film up to that point suggesting aliens don't exist. We submit that it is a conflict of this sort—or better, *tension*—that is really the issue when it comes to Christian belief and the existence of ETI. The existence of ETI would come as a major shock to Christians, because heretofore ETI simply didn't enter into the story about life, the universe, and everything that they took to be complete. ETI would, at the very least, entail that that story was radically incomplete. More extreme, the existence of ETI might cause some to think we're living in an entirely different story altogether. It would be akin to either a major unforeseen plot twist, or swapping one book for another one.

(McIntosh and McNabb, 2021, p. 13)

McIntosh and McNabb then go onto to resolve this dilemma by examining a number of works of fiction which involve both alien life in S4 and S5 scenarios and a Christian narrative. (They focus on S4 and S5 scenarios because they are the most challenging scenarios. As a result, so they argue, if they can be resolved, we are justified in assuming that S1, S2, and S3 scenarios can also be resolved.)

I want to add to McIntosh and McNabb's argument by agreeing that, at least at first, the discovery of intelligent alien life might cause Christians to question the narrative of Christianity, but this is entirely to be expected because of a consistent pattern of subverted expectations that can be found in both the Old and New Testaments. In each case, as we shall see, expectations are subverted, sometimes in a distressing way, but in the process, a grander more complete narrative and understanding of God and his relationship to humanity emerges. As a result, Christians can acknowledge that the discovery of intelligent alien life might at first seem to conflict with our (current) narrative expectations of the world, whilst also incorporating this into a broader Christian (meta-) narrative, namely that the Christian story is one of subverted expectations and that the unexpected discovery of alien life in the long run will lead to a grander and more complete understanding of God and his relationship to his creation. I will illustrate this claim with a few examples.

The Least Likely Person for the Job

One theme that consistently emerges throughout the Old Testament is that God often works through unexpected and unlikely people. For example, Moses, a murderer who was 'slow of speech' (Exodus 4:10) is chosen by God to lead his people out of Egypt (Exodus 3:1–22). Gideon who was the weakest member of the weakest clan in Manasseh is greeted by God's angel as a 'mighty warrior' who is then chosen to free the Israelites from Midianite oppression (Judges 6–7). David, a shepherd boy and youngest of his brothers was selected by God to become the King of Israel (1 Samuel 16:1–13). This same David is also able to defeat the giant Goliath despite being armed only with a sling and with no armour (1 Samuel 17). David is also declared a 'man after God's own heart' (1 Samuel 13:14, Acts 13:22), despite the fact that he slept with one of his soldier's wives and then arranged to have him killed to cover up the resulting pregnancy (2 Samuel 11).

Another series of particularly striking illustrations concerns God's use of women in the intensely patriarchal ancient Near East. For example, Esther, a young Jewish girl, becomes Queen of Persia and saves her people from annihilation in the book of Esther (Esther 2:1–18), and, in the Book of Judges, Deborah is selected to lead Israel against the Canaanites (Judges 4–5). Arguably she is the most successful of all the judges!

Countless other examples could be given, but at least one of the points we are supposed to draw from these subverted human expectations is that God 'does not look at the things people look at' (1 Samuel 16:7) and that he can bring strength out of weakness so that the glory and honour are his alone since the people he chooses could not do the things they end up doing without him (1 Corinthians 1:27–29). Hopefully, for our purposes, the point is clear. In each example, we have a subverted expectation, but out of this subverted expectation comes a more complete picture of God.

It is Always Darkest Before the Dawn

Another theme that consistently emerges is that God will create something good, human beings will completely ruin it resulting in disaster, only for God to bring something good out of that disaster. We see this right at the start of the Old Testament. God creates the world and it is 'very good' (Genesis 1:31), but by end of Genesis 3, human beings have already ruined everything: Adam and Eve have eaten the fruit of the tree of the knowledge of good and evil and are banished from the Garden of Eden. As a result, they are made to endure the sufferings, struggles, and evils of the world as we know it. Despite this, God does not abandon them. For example, he creates clothes for them (Genesis 3:21), and he supports Eve through childbirth (Genesis 4:1). More generally, God continues to provide for human beings despite their fallen nature (Philippians 4:19), and, ultimately, he redeems all of humanity through Jesus's death on the cross (John 3:16–17, Romans 5:6–19).

Another example involves the life of the Old Testament figure Joseph. Who, due to envy, is sold into slavery by his brothers and who is then taken away from his father and into Egypt. Despite this, however, God raises Joseph up to become the vizier or prime minister of Egypt, second only to Pharoah. In this role, Joseph is able to guide the Egyptians through a difficult famine. Eventually, he is reconciled with his brothers and reunited with his father (Genesis 37–50). Again, the pattern is clear: human beings ruin thing due to sin and ineptitude, but God finds a way to bring something good out of that disaster.

Another example involves the Kingdom of Israel as a whole. God chooses the Jewish people, and Israel as a nation, to be his representatives on Earth (Deuteronomy 7:6–11). He sets up a covenant with them, promising to provide for them as long as they follow his commandments, but consistently they fail to do this causing the Kingdom of Israel to split into two. The northern kingdom of Israel is eventually destroyed by the Assyrians (2 Kings 17:5–8). Sometime after this, the southern kingdom of Judah is also invaded, this time by the Babylonians, the Temple is destroyed, and the people are forced into exile (2 Kings 25:8–12). Despite this, even though he has no obligation to do so (given Israel and Judah broke their covenant with him), God preserves a remnant (Isaiah 65:9) and eventually, through the Persian king Cyrus, restores them to Judah (Ezrah 1:2–3).

The pattern emerges yet again. We might expect, for example, that God's chosen people would always obey him (Why else would he choose them if they weren't going to obey him?) and that they would flourish as a result. We might also expect that once human beings have disobeyed God, he would (justifiably) wash his hands of them. However, this isn't what we see. Instead, we see that human beings consistently disobey God even though he has been good to them. This disobedience results in ruin and chaos, but despite this, God remains faithful to humanity and he then finds a way to bring something good out of these man-made disasters. Once again, our expectations are subverted.

Jesus of Nazareth

The examples given so far of the ways in which God subverts human expectations are all taken from the Old Testament and thus are equally available to both Jews and Christians. However, for Christians, the most powerful example, which in many ways contains within it many of the patterns we have just explored, is that of the life, death, and resurrection of Jesus of Nazareth.

For a start, God uses a woman to bring Jesus into the world (Luke 1:26–38), another example of God using women in the patriarchal ancient Near East. The Jewish people are expecting a political and military Messiah, but instead Jesus is from a modest background and leads a peaceful ministry. The Jewish people expect him to overthrow their enemies, but instead Jesus is killed by them (Matthew 27:32–56). His enemies think that this is the end of the matter, whilst his followers despair, only for Jesus to be resurrected from the dead three days later, thus paving the way for all human beings to be saved (Luke 24, Colossians 1:15–20, Romans 5:10–11). Finally, the Jewish people are expecting a merely human Messiah. Of course, Jesus is fully human, but he is also God incarnate and the Second Person of the Trinity (John 1:1–18).

In Science Fiction

Numerous other Biblical examples could be given but hopefully the point is clear. The Christian and Jewish story is one of subverted expectations and narrative tension. Human beings expect one thing, and then something quite unexpected happens. The only thing we can expect God to do is the unexpected!

We actually see this happening in various examples of science fiction. For example, in 'The Fire Balloons', discussed earlier, the priests are initially surprised by the response of the Martians, but they soon come to see that they are being invited into grander narrative. Another example, this time imagining a negative interaction between humans and extraterrestrials (an S4 scenario as explained earlier), is explored in Mary Doria Russell's *The Sparrow* (1996).[5] In it a group of human beings led by some Jesuit missionaries attempt to make contact with an alien species. All but one of them is killed, many of them by the aliens themselves. The sole survivor, a priest, after being mutilated and tortured by the aliens, returns home. Understandably, he is confused and angry at God and the book ends with him trying to make sense of it all. Whilst I, of course, do not want to make light of this fictional character's suffering, as a reader we can see that various events have been set in motion by this failed expedition which, once they have come to fruition, may allow us (and him!) to make sense of the suffering described in the book. (There is a sequel, *Children of God* (1998), which explicitly spells out how some of these events unfold, but for our purposes the point is that even in a fictional S4 scenario with a theologically ambiguous ending, the possibility of resolving the narrative tension within a Christian framework is there.)

We can confidently conclude, therefore, that Christians should not fear the possibility of narrative tension should we discover intelligent alien life in the future. If Christianity is correct, such tension will be reconciled either in the sort run (as in 'The Fire Balloons') or in the long run (as in *The Sparrow*), and when it is a more complete and glorious picture will emerge!

'The heavens declare the glory of God' (Psalm 19:1)

So far in this chapter we have seen a number of things. First, we saw that some thinkers have held that the discovery of extraterrestrial life would pose a problem for adherents of the Abrahamic faiths. However, we then saw that theologians have been carefully thinking about these issues since the medieval period and that these problems are largely illusory. I then laid out a general 'one-size-fits-all' strategy to deal with extraterrestrials, namely that there is nothing to stop us from viewing Judaism, Christianity, or Islam (depending upon our commitments) as an earthly religion for earthly human beings. This, then, gives us endless flexibility when thinking about how aliens might fit into a Cosmic Judaism, Christianity, or Islam.

In the second section, the issue of subverted expectations was examined. By way of response, I showed that subverted explanations are part of both the Jewish tradition in the Old Testament and also particularly in the Christian tradition in the life, death, and resurrection of Jesus of Nazareth in the New Testament. I concluded that even though the discovery of alien life might subvert our theological narrative expectations, Jews and Christians can be confident that in the long run, these tensions will be resolved and that a grander more complete understanding of God will emerge. (Such a strategy may also be available to Muslims should similar themes be identified within their tradition. Whether these themes are there to be found, however, I shall leave to Muslim theologians.[6])

Up until this point, this chapter has largely been theological in nature, but I want to end on a philosophical point. Given their scriptures are silent on the issue, are there any philosophical reasons why adherents of the Abrahamic faiths might expect there to be extraterrestrial life? (Given all three religions believe in the God of classical theism, all three traditions can be treated as a whole.)

I believe that there are some reasons to think it likely that there is life, perhaps even of an intelligent sort, out there. We can see this by considering why God created the Universe in the first place. For the Abrahamic faiths, God is under no obligation to create the Universe, nor does he create it to fulfil some need of his. This is because God is perfect in and of himself. He has no needs and is under no obligations. Thus, the reason God created the Universe is simply as a voluntary manifestation of his goodness, power, and love.

We can analyze this in further detail by considering Aquinas's description of God as pure act (Aquinas, ST 1, Q12, A1, co.) or existence itself (Aquinas, ST 1, Q3, A4, co. and Q4, A2, co.). Thus, everything that exists partakes in the nature of God in as much as it exists. By the same token, everything good partakes in the nature of God because of his perfect goodness in as much as it is good. Likewise, everything beautiful partakes in the nature of God in as much as it is beautiful, and so on (Aquinas, ST 1, Q2, A3, co.). As a result, the entire Universe expresses the divine characteristics.

Here it is worth considering another of God's attributes. According to the Abrahamic traditions, God is infinite, timeless, and boundless (Aquinas, ST 1, Q7, A1). A number of interesting things follow from these various observations and insights.

First, the size of the Universe and the diversity of life here on Earth now seems much more likely given theism. Given that the Universe, and life here on Earth, expresses the divine characteristics, and given God is infinite, the size of the Universe and the vast diversity of life on Earth make sense, since it is only through this vast size and diversity that God's infinite nature can be (imperfectly) expressed. Created things can never fully express all of God's characteristics, but a vast diversity of created things, all differing in their various attributes, can more fully express more of God's characteristics, and in so doing, they each derive value and worth.

Second, following on from this, it shouldn't surprise us if we find life elsewhere in the Universe. God is infinite, and the Universe and the life within it are expressions of his nature. As a result, it shouldn't surprise us to find life elsewhere in the Universe, and it shouldn't surprise us if it looks very different to us. All of it will, in some way, express some aspect of God's nature and through studying it and interacting with it, we can learn more about God himself.

Finally, it is worth returning to the question of human uniqueness. As we saw, Davies (2010) stated that 'any theology with an insistence on human uniqueness would be doomed' (p. 193). However, it now becomes entirely unclear why that should be the case. Why should human uniqueness be threatened by the discovery of intelligent alien life? Sure enough, its discovery would show us that we are not unique in our intelligence, but it is unclear why this should disturb us. There are many different ways we can be unique and if, as I have argued, every living creature uniquely expresses some aspect of God, then it follows that we too uniquely express some aspect of God's nature. This holds even if there are other intelligent alien life forms out there which also uniquely express some aspect of God's nature. This can be seen by considering the fact that there are roughly eight billion human beings on the planet, but each one of us is unique. Likewise, even if there are eight billion intelligent species scattered across the Universe, this does nothing to demonstrate that the human species is not unique as a species, and that you are not unique as an individual!

Conclusion

In this chapter I have done a number of things. I began by showing that some thinkers have held that the discovery of alien life would pose a problem to the Abrahamic faiths. I then demonstrated that theologians have not been idle in this regard and have been carefully thinking about these issues since the medieval period. In the process I laid out some of the key issues such (astro)theologians have identified. I directed the reader towards various sources which engage with those issues whilst also proposing a 'one-size-fits-all' response to such issues. I then honed in on the issue of narrative tension. I argued that for Christians, and to a lesser extent Jews, narrative tension and subverted expectations are part of their theological worldview and that, as a result, even if initially the discovery of alien life were to generate narrative tension, they need not worry as such tension will be resolved in the long run and that in the process a grander and more complete picture of God will emerge. Finally, I put forward a philosophical argument for why, if the Abrahamic God exists, it should not surprise us if there is alien life out there. I conclude that Jews, Christians, and Muslims can enthusiastically engage in the search for extraterrestrial life and need not fear the implications of its discovery for their respective traditions.

Perhaps we will never find it. If so, it may be that it isn't out there. Alternatively, it may be that God in his providential wisdom has separated us by such vast distances that we will never discover them, perhaps to protect us from them or them from us or perhaps for some other reason. (I suppose, if we never find alien life, we'll never know which hypothesis is correct: Is it not out there? Or is it simply located in the next galaxy along which we haven't yet explored? There may well be no way to choose between these too options.)

However, if we do find alien life out there, particularly of an intelligent sort, followers of the Abrahamic faiths should remember that it, like us, expresses something of the Divine. We must therefore remember to treat it accordingly.

Notes

1 Arguably, the closest we come to this is in Qur'an 7:54. In it, Allah is described as 'Lord of the worlds'. The plural 'worlds' could be seen as indicative of other inhabited planets in the Universe, and no doubt if alien life is discovered, Muslims will interpret it as such. However, in and of itself, this description of God is silent on the question of alien life, since 'worlds' could be referring to the various spiritual and, at least one, physical worlds that make up Muslim cosmology, for example, the invisible world of the jinn and the visible world of human beings. As a result, this passage is of little help in this context.
2 All of the sources just referenced, when discussing Islam, except for Playford, Bullivant, and Siefert (2024), can be found in Malik and Determann (eds.) (2024), which, as far as I am aware, is currently the most in-depth (and recent) discussion of Islam and extraterrestrial life.
3 Technically, Ratramnus did consider the dog-headed men to be humans because he subscribed to Aristotle's definition of man as a rational animal, and he believed, based on reports available to him at the time, that the *cynocephali* were rational. Clearly, however, dog-headed men are not

conventionally human even though they share with us a rational and animal nature. In this sense, they mirror extraterrestrial intelligences who, should they exist, will also share with us a rational and animal nature, even if they manifest that nature in a very different way to us and are very different to us in other respects. See Bruce (2006) and Matter (2006) for a more detailed discussion of Ratramnus's views on the dog-headed men.

4   In particular, they discuss the relationship between sin and the body. They consider the possibility that the different ways we can sin are limited by our bodies and that different things may sinful for aliens with different bodies.

5   McIntosh and McNabb (2021) also discuss *The Sparrow*, along with various other works of fiction.

6   It is interesting that the Prophet Muhammad is usually considered to have been illiterate. This make him an odd choice, by human standards, to be the vehicle through which Allah revealed the Qur'an. Perhaps this is a Muslim example of subverted expectations.

References

Abdullah, Faisal (2024) 'Classical Muslim Thought and the Theological Implications and Possibility of Non-Human Entities Bearing Higher Intelligence', in *Islamic Theology and Extraterrestrial Life: New Frontiers in Science and Religion*, edited by Shoaib Ahmed Malik and Jörg Matthias Determann. London: I.B. Tauris, pp. 115–138.

Aquinas, Thomas (1948) *Summa Theologica*, translated by the Fathers of the English Dominican Province. New York: Benziger Bros.

Bradbury, Ray (2008) *The Martian Chronicles*. London: Harper Collins.

Bruce, Scott G. (2006) 'Hagiography as Monstrous Ethnography: A Note on Ratramnus of Corbie's Letter Concerning the Conversion of the Cynocephali', in *Insignis Sophiae Arcator: Medieval Latin Studies in Honour of Michael Herren on his 65th Birthday*, edited by Gernot Wieland, Carin Ruff, and Ross G. Arthur. Turnhout: Brepols Publishers, pp. 45–56.

D'Costa, Gavin (2009) *Theology and World Religions: Disputed Questions in the Theology of Religions*. Chichester: Wiley-Blackwell.

Davies, Paul. C. W., (2010) *The Eerie Silence: Renewing our Search for Alien Intelligence*. New York: Houghton Mifflin Harcourt.

Davison, Andrew (2023) *Astrobiology and Christian Doctrine: Exploring the implications of life in the universe*. Cambridge: Cambridge University Press.

Foltz, Bruce (2019) *Medieval Philosophy: A Multicultural Reader*. London: Bloomsbury.

Hick, John (2004) *An Interpretation of Religion: Human Responses to the Transcendent*, 2nd edition. New Haven: Yale University Press.

The Holy Bible (1979) New International Version. London: Hodder & Stoughton.

Ioannides, Mara Wendy Cohen (2019) 'Judaism and Extraterrestrials: Theological Lessons from Science Fiction', *The Journal of Popular Culture*, Vol. 52, No. 5, pp. 1200–1217.

Jalajel, David Solomon (2024) 'Extraterrestrials and Moral Accountability: Nonhuman Moral Personhood through the Lens of Classical Sunni Theology and Law', in *Islamic Theology and Extraterrestrial Life: New Frontiers in Science and Religion*, edited by Shoaib Ahmed Malik and Jörg Matthias Determann. London: I.B. Tauris, pp. 87–113.

Jha, Alok (2010) 'Pope's Astronomer Says He Would Baptise an Alien If It Asked Him', *The Guardian*, 17 September, online. Available at: www.theguardian.com/science/2010/sep/17/pope-astronomer-baptise-aliens Accessed 23 June 2024.

Karamali, Hamza (2024) 'Theological Information on the Existence of Intelligent Life Outside Our Solar System: Metaphysics, Scripture, and Science', in *Islamic Theology and Extraterrestrial Life: New Frontiers in Science and Religion*, edited by Shoaib Ahmed Malik and Jörg Matthias Determann. London: I.B. Tauris, pp. 25–41.

Lamm, Norman (1965) 'The Religious Implications of Extraterrestrial Life', *Tradition: A Journal of Orthodox Jewish Thought*, Vol. 7, No. 8, pp. 5–56.

Malik, Shoaib Ahmed and Determann, Jörg Matthias (eds.) (2024) *Islamic Theology and Extraterrestrial Life: New Frontiers in Science and Religion*. London: I.B. Tauris.

Matter, Ann E. (2006) 'The Soul of the Dog-Man: Ratramnus of Corbie between Theology and Philosophy', *Rivista di Storia della Filosofia*, Vol. 61, No. 1, Filosofie e Teologie, pp. 43–53.

McColley, Grant and Miller, H. W. (1937) 'Saint Bonaventure, Francis Mayron, William Vorilong, and the Doctrine of a Plurality of Worlds', *Speculum*, Vol. 12, No. 3, pp. 386–389.

McIntosh, C. A. and McNabb, Tyler Dalton (2021) 'Houston, Do We Have a Problem? Extraterrestrial Intelligent Life and Christian Belief', *Philosophia Christi*, Vol. 23, No. 1, pp. 101–124.

Montasseri, Mohammad Mahdi (2024) 'Islamic Sacred Resources on Extraterrestrials and Their Possible Eschatological Implications', in *Islamic Theology and Extraterrestrial Life: New Frontiers in Science and Religion*, edited by Shoaib Ahmed Malik and Jörg Matthias Determann. London: I.B. Tauris, pp. 59–86.

Paine, Thomas (1794) *The Age of Reason: Being an Investigation of True and Fabulous Theology*. Paris: Barois.

Peters, T. (2011). 'The Implications of the Discovery of Extra-terrestrial Life for Religion', *Philosophical Transactions: Mathematical, Physical and Engineering Sciences*, Vol. 369, No. 1936, pp. 644–655.

Playford, Richard (2024) 'The Alien in the Lamp? The Djinn and Alien Life in Islamic Theology', in *Islamic Theology and Extraterrestrial Life: New Frontiers in Science and Religion*, edited by Shoaib Ahmed Malik and Jörg Matthias Determann. London: I.B. Tauris, pp. 159–173.

The Qur'an: Sahih International Translation (1997) Translated by Umm Muhammad, Mary Kennedy and Amatullah Bantley. Saudi Arabia: Dar Abul Qasim.

Russell, Mary Doria (1996) *The Sparrow*. New York: Villard Books.

Russell, Mary Doria (1998) *Children of God*. New York: Villard Books.

10

IF NOT GOD, THEN...ALIENS?

The Ancient Astronaut Hypothesis and Popular Atheism

Stephen Bullivant

Introduction

The purpose of this chapter is quite simple. It is perhaps best described as an exercise of 'philosophy of religion *out in the wild*'. That is to say, I seek here to explicate the main contours of a popular and culturally significant body of exophilosophical argumentation which, for the most part, exists outside of the realm of peer-reviewed articles and academic monographs: the Ancient Astronaut Hypothesis (AAH), sometimes known more informally as 'ancient aliens theory'. (Both terms are used by its protagonists; for brevity's sake I'll normally use 'AAH' herein). As shall be detailed below, the AAH appears in many forms. Here, however, I shall mainly focus on what is arguably its most culturally prevalent form, whose *loci classici* include the works of million-selling authors Erich von Däniken (1935–) and Zecharia Sitchin (1920–2010), and the History Channel's long-running documentary series *Ancient Aliens* (2009–). This form of AAH, I will argue, might fairly be taxonomized as a curious species of *scripturally literalist, Hickian, technological Euhemerism* (!). The extent to which it is also *atheistic*, which is a more complicated one, will be discussed at some length.

In what follows, I first introduce the basic tenets of the AAH. I then sketch its appearance in three distinct spheres: science fiction, the belief systems of a number of explicitly religious groups, and the 'alternative archaeology/history' milieu. The focus of this chapter – what I will, for convenience's sake, be calling 'mainstream AAH' – is principally a variant of the latter group (albeit with significant cross-fertilization with science fiction). After some background and context, I will explain each of the italicized 'hallmarks' of mainstream AAH (albeit not in the order given earlier), showing how it is 'technological(ly) Euhemerist', 'scripturally literalist', 'Hickian', and (in practice, but not necessarily) 'atheistic'. What I mean by each of these terms will, of course, be explained at the relevant juncture.

DOI: 10.4324/9781003440130-15

The Ancient Alien Hypothesis

Broadly speaking, the AAH posits highly advanced extraterrestrials having visited Earth in the remote past (and perhaps on an ongoing basis ever since) as the best means of explaining a wide range of phenomena, including some or all of:

- Reports of supernatural beings (most obviously gods, but also angels, demons, jinns, magical creatures, etc.), or extraordinary events or objects in ancient literature, especially in overtly religious or mythological writings (though not exclusively: e.g., Plato's *Timaeus* and *Critias* regarding Atlantis);
- Depictions of the above, and/or seemingly anachronistic elements (e.g., an alleged 'rocket ship' on a seventh-century Mayan funerary relief), in ancient art;
- Certain items of ancient material culture, especially monumental architecture or megaliths (e.g., pyramids of Giza, Easter Island/Rapa Nui *moai*) but also intricate or unusual smaller objects, whose purpose and/or construction is (allegedly) difficult to explain in conventional terms;
- Either the existence, or particular capabilities (e.g., high intelligence), of human beings.

More specific examples will be given in what follows. Here, suffice it to say that AAH arguments are typically eclectic and cumulative, offering a most plausible, 'best fit' explanation for a large number of otherwise puzzling or anomalous phenomena. No single piece of evidence is ever accorded too much weight in the overall schema. In practice, this makes AAH arguments very difficult to argue against. While many scholars – scientists, linguists, archaeologists, historians – have debunked many specific claims made by AAH authors over the years (e.g., Story, 1976; Fagan, 2006; Colavito, 2020), this takes up a great deal of time and energy. Meanwhile, AAH theorists can simply admit that this or that claim of theirs was indeed 'Total poppycock!' (von Däniken [1999] 2019), but then point to all the *other* still-undebunked claims. For example, although accepting research proving that the remarkable geoglyphs in Nazcar, Mexico were not, *contra* one of von Däniken's signature theories (made in multiple books over the decades, e.g., [1968] 2019, 1998), an airport for alien craft, fellow AAH theorist Philip Coppens writes:

> On rare occasions such as this, science has addressed the Ancient Alien Question, but still, at more than 40 years later, most of the 230 questions posed by von Däniken [in *Chariots of the Gods?*] remain unanswered by science. Worse, science refuses to pose the questions itself, and, almost half a century later, it therefore remains the task of people outside the scientific community to ask the question again.
>
> *(Coppens, 2011, p. 4)*

Two things are worth highlighting here. First, the rhetorical strategy of 'only asking questions' is a common one in AAH literature ('Could it be that…?', 'Is it conceivable….?') and allows for an easy get out when the answers are negative (e.g., 'here's what the critics overlooked: […] Questions are the opposite of assertions'; von Däniken [1999] 2019). Second, note the idea that AAH theorists have had to take matters into their own hands due to the (perceived) dereliction of duty from 'the professionals'. This taking pride in a maverick status is common in other 'fringe science' (Gordin, 2021, pp. 54–55) subcultures, such as astrology and cryptozoology, too. By explicitly taking into account 'data that Science has excluded' (Fort, 1919, p. 1), or 'certain ancient myths and traditions, judged to be of no historic value by scholars' (Hancock, 2015, p. 47), or points of view excluded by the 'thought prison' of 'the wise academics of our age' (von Däniken, 2009, p. 78), one can claim to be the true, and truly sceptical, scientist or historian.

There are three main 'places' where one may encounter variants of the AAH. The first is in science fiction. The basic idea of aliens having interacted, in multiple possible ways, with human beings in the remote past has proven irresistible to authors and screenwriters. Notable literary examples include H. P. Lovecraft's 'The Call of Cthulu' (1928), and Arthur C. Clarke's *Childhood's End* (1953) and *2001: A Space Odyssey* (1968). Movies range from Stanley Kubrick's version of the latter (also 1968), to *Stargate* (1994), to *Indiana Jones and the Kingdom of the Crystal Skull* (2008). In television, the original *Stargate* movie spawned several series (*SG-1*, 1997–2007; *Atlantis*, 2004–2009; *Universe*, 2009–2011). Various iterations of *Star Trek*, including the *Original Series* (1966–1969), *Deep Space 9* (1993–1999), and *Voyager* (1995–2001), played with the idea in different ways (e.g., Asa, 1999; Linford, 1999).

The second is in a diverse array of new religious movements (NRMs). For example, the idea that great religious figures of the past – e.g., Jesus, Buddha, Laozi, Krishna, Isis, Osiris, Muhammad – either were themselves, or else in contact with, advanced extraterrestrials has been claimed by groups, including Heaven's Gate (Zeller, 2014), the Seekers (Festinger et al., 1956), the Unarius Academy of Science (Tumminia, 2005), and the Aetherius Society (Rothstein, 2021). Other groups incorporate aliens into elaborate accounts of Creation, as in Scientology's 'esoteric, space opera cosmology' (Urban, 2021, p. 336; see also Urban, 2011), or Raëlism's interpretation of the biblical 'Elohim' – a common term for God in the Hebrew Bible – as a race of powerful aliens (Palmer, 2004; Gallagher, 2010; Dericquebourg, 2021).

The third is, as noted earlier, from within what is perhaps best described as the 'alternative archaeology/history milieu', of which AAH-specific literature is but a subset. Works in this sphere position themselves as *non-fiction* books or documentaries (i.e., not as science fiction), and as serious, scholarly scientific or historical enquiries 'following the argument wherever it leads', rather than as part of any religious or spiritual package. This possibility is sometimes specifically denied:

> All of this has nothing to do with religion. I will turn myself around in my grave if my ideas turn into a cult. It is not about belief – it is about testing. The facts are there.
>
> *(von Däniken, 2011)*

It is this strand of AAH speculation that is the most culturally well-known, hence my terming it 'mainstream AAH'. Though not the first work to explore these ideas, the Swiss author Erich von Däniken's 1968 book *Erinnerungen an die Zukunft*, literally 'Memories of the Future' but published in English as *Chariots of the Gods?*, is the landmark text here. A huge, international bestseller in its own right, it also spawned a 1970 movie, which received a 'Best Documentary' Oscar nomination. Since then, von Däniken has published dozens of other books, all (re)probing aspects of the same basic idea, with reported book sales of over 70 million.

Also important here are the (likewise millions-selling) works of the Azerbaijani-born American writer Zecharia Sitchin, whose debut *Twelfth Planet* was published in 1976. These collectively present a detailed metanarrative, largely missing from von Däniken's works up to then, but which he thereafter incorporated, and which appears frequently in other AAH works. Sitchin's schema is purportedly based on documentary evidence from ancient Mesopotamia. Put briefly, it posits the existence of a race of aliens – the Annunaki (also identified with the Biblical *Nephilim* and *Elohim*) – who originally came to Earth in the remote past to mine gold. Tiring of their labour, they created a new set of beings to do the dirty work for them by hybridizing themselves with *Homo erectus*. *Homo sapiens* is thus 'a slave made to order' (1990, p. 158). The Annunaki stayed around for several centuries, ruling over humans as 'God-kings', and using their labour (combined with the Annunaki's own advanced technological skills) to build structures such as the Pyramids.

The ideas of both von Däniken and Sitchin are featured prominently in the History channel's hit documentary series *Ancient Aliens*, initially commissioned as a two-part special in 2009, but still going strong in its nineteenth season, with 240 episodes in the can. Its Executive Producer, and most recognizable (and much-memed) 'talking head', Giorgio Tsoukalos, was mentored by von Däniken, who appears in many of the episodes. The show has engendered a companion volume (Producers of Ancient Aliens, 2016), a computer game with a very Sitchin-esque plot, and occasional 'Alien Con' events attracting thousands of attendees. A feature film is currently in the works.

From the above, it is clear that 'mainstream AAH' constitutes a significant social and cultural phenomenon. Quite how tens of millions of books sold in multiple languages, and one of America's (if not the world's) most popular documentary series, translate into actual believers/adherents of AAH ideas is, admittedly, difficult to judge. The types of large-scale social surveys needed to find out have not been done. However, when surveys *have* been done in the USA and elsewhere, asking about related paranormal/alternative beliefs – UFOs, Bigfoot, ghosts, etc. – one finds that decent proportions of the public do indeed hold them (Bader et al., 2017; Bullivant et al., 2019).

Although internally diverse, mainstream AAH coalesces around a quite distinctive set of philosophical positions. The rest of this chapter will explore these.

Technological Euhemerism

Euhemerus of Messene (modern-day Messina in Sicily) was a Greek writer of the late-fourth to early-third centuries BC. His main work, *The Sacred History*, is now lost, but substantial fragments are quoted by later writers, including Diodorus Siculus and Eusebius of Caesarea. Judged by these, the work appears to be a philosophical romance, built around a fictional travelogue to far-distant lands. In it, Euhemerus describes visiting a temple on an island in the Arabian Sea, whose priests claim to be descended from Cretan immigrants of the remote past. The temple contains an inscription, written in Cretan, chronicling the early history of the royal house of Crete. These are, it turns out, the figures revered as gods and goddesses by the Greeks of Euhemerus' day:

> Uranus was the first king, a benefactor to man and versed in the movements of the stars... Hestia bore him two sons and two daughters, Titan, Cronus, Rhea and Demeter.

Cronus followed his father, and married his sister Rhea, by whom he had Zeus, Hera and Poseidon....

(Quoted from Brown, 1946, p. 261)

Euhemerus' point, evidently, is that the gods were originally mortals, whose stories have grown, gradually but greatly, in the telling. It is perhaps no coincidence that he was writing in the decades following the death of Alexander of the Great, who was already being revered as a god in much of his former Empire, and to whose already-impressive deeds, a steady stream of new legends were being added. With that in mind, note especially Euhemerus' following comment: 'Zeus was a great conqueror, visiting almost all lands and being universally acknowledged as a god' (ibid.).

Following its eponym, 'Euhemerist' accounts of religious or mythological beliefs offer a naturalistic aetiology of how those beliefs arose and came to be believed. Real, flesh-and-blood individuals or events are posited as constituting an historical hard core. However, over time, they have been elaborated and apotheosized, either deliberately (i.e., by those who stand to benefit, such as the claimed descendants of the original figures), or else by a more organic exaggeration-over-time. Broadly Euhemerist accounts have been influential tools for explaining (away) religious and mythological beliefs and stories, albeit often now supplemented or supplanted by others drawing more on the disciplines of psychology, sociology, and anthropology (compare, e.g., the classic accounts of Hume, Freud, Marx, and Durkheim; or more recently, see Boyer, 2001).

Our AAH proponents stick more closely to a classic Euhemerist approach, albeit with two special twists. Most obviously, of course, Euhemerus' 'ancient royals' are replaced by ancient aliens. But they are also, critically, ancient aliens with futuristic technology. Hence as Tsoukalos puts it programmatically in the very first episode of *Ancient Aliens*:

Whatever was described in the Old Testament wasn't 'God'. It was a misunderstood flesh-and-blood Extraterrestrial whom our ancestors misinterpreted as being divine and supernatural. Why? Because of misunderstood technology. And that's the thread that applies to all of the Ancient Astronaut Theory.

(Ancient Aliens, 2010a)

Or according to the Tsoukalos avatar in the franchise's official computer game: 'I say these "gods" were just flesh and blood space travellers, misinterpreted as divine creatures because of the technology they used'.

Technology is a major strand of aetiological explanation, often receiving more direct attention than the aliens themselves (see, e.g., Childress, 2000). The most obvious example is present in the title of *Chariots of the Gods?* itself. Any and all mentions of flying machines (Elijah's 'chariots of fire' [2 Kings 2.11], flying carpets in the Arabian nights, *vimānas* in the Vedas [see Long's chapter in this volume]), flying creatures (dragons in Chinese mythology, Native American thunder birds, Muhammad's horse *Buraqa* for his famous night journey from Mecca to Jerusalem), or special meteorological phenomena (the dancing Sun at Fatima in 1917, the 'pillar of the cloud' in Exodus, 'the cloud' at Jesus's ascension [Acts 1.9]) can be satisfactorily explained as alien spacecraft of one sort or another. But this is only the tip of the iceberg. According to mainstream AAH's technological Euhemerist hermeneutic, Zeus and his lightning bolt become an alien warlord with a 'directed energy beam weapon'

(Ancient Aliens, 2010b). The Great Pyramid of Giza was a power plant (Dunn, 1998; Ancient Aliens, 2023). And the Ark of the Covenant was one or more of (i) a factory for edible algae ('manna') that needed cleaning once a week ('the Sabbath') (Ancient Aliens, 2010a, 2019); (ii) a radio set so that Moses could communicate with an alien ('the Lord') in the spacecraft ('the pillar of the cloud'), accompanying the Hebrews in the desert (von Däniken [1968] 2019, p. 50); or (iii) 'a modified Tesla coil that spewed out deadly fireballs in every direction around the device' (Childress, 2016, p. 77).

It is worth stressing here that, whatever else they might be, these are still straightforwardly naturalistic explanations. The impossibility of supernatural explanations is a basic (though rarely if ever argued for) assumption, which is taken to point to the *only other* plausible explanation: flesh-and-blood extraterrestrials with nifty gadgets. Hence:

> There's only two possibilities – either God did it, which we really don't think happened, or some hi-tech civilization from another planet came....
>
> *(Tsoukalos, in Ancient Aliens, 2010a)*

> We have these stories of these 'gods' that have these supernatural, magical powers. Let's be honest, magical, supernatural powers don't really exist. So what was it that was described? In my opinion, it was people who had access to advanced technology.
>
> *(Tsoukalos, in Ancient Aliens, 2014)*

> [Liaisons between 'gods' and women report in mythology must in fact be] with extraterrestrials, not with gods, because gods do not exist.
>
> *(Tsoukalos, in Ancient Aliens, 2010b)*

Scriptural Literalism

Aliens aside, AAH proponents adopt an otherwise naively literal approach to scriptural and mythological texts. There is, for instance, never any engagement with questions of dating, authorship, provenance, or genre – that is to say, the stock in trade of modern scripture scholars or mythographers. Rather, texts or oral tradition are taken as more-or-less straightforward, reliable reportage, albeit filtered through the eyes and thought categories of primitive, pre-technological people. For example, if Exodus describes the Red Sea being parted, then the Red Sea *was* parted, and there must, therefore, have been a being that both wanted to part it and was able to do so. With God ruled out *a priori*, that leaves the path clear for something else sufficiently impressive and powerful – i.e., hi-tech aliens – that the ancient Israelites were only able to (mis)interpret in terms of a divine being. (Note that the possibilities that the Red Sea was not really parted, or that Exodus was written centuries after the events it purports to describe, are nowhere considered.) This same hermeneutical approach can be applied, *mutatis mutandis*, more or less infinitely (hence 240 episodes and counting).

Another telling example of this literalist, face-value approach to religious and mythological texts comes in AAH interpretations of human origins and historical development. Traditional beliefs, found in many different cultures, that human beings are descended from gods, or were made in 'God's image', are likewise easily incorporated into the AAH. We noted earlier Zecharia Sitchin's belief that members of *Homo sapiens* were bioengineered by the Annunaki, initially as gold-mining slaves, though over the centuries were guided and educated – 'civilized', indeed – to develop their potential. Details may differ, but this basic idea has caught on in a big way among AAH enthusiasts. Alleged 'proofs' of alien intervention in the human genetic code (von Däniken, 2017) and human culture – religion, politics, law, technology – abound. Evidently, there are analogies to be drawn here to some modern Creationist and/or Intelligent Design accounts of human origins. And along with this also comes an AAH retooling of fairly strong accounts of divine providence (cf. Ps 8.4–6):

> We've always been their 'pet project', so to speak, meaning that they helped us to become who we are today.
>
> *(Tsoukalos, in Ancient Aliens, 2013b)*

> You have to wonder… why do they care about us? Why do they want to help us, give us prophecies, instruct us in better lives? In many ways, it is only natural that they would. We're like their children.
>
> *(Childress, in Ancient Aliens. 2013b)*

Hickian

In characterizing the AAH as 'Hickian', I am, of course, referring to the British philosopher of religion John Hick (1922–2012) and, in particular, to his influential theory of religious pluralism. Very simply put, this posits that each of the major world religions is an authentic, if partial and culturally influenced, response to the same 'transcendent reality'. Hence within the 'bewildering plurality' of seemingly incompatible truth-claims, there is a real core to which all are bearing genuine witness (Hick, 2004, p. 2). All are therefore 'true' and legitimate paths converging on the same destination. The details, or indeed cogency, of Hick's religious 'meta-theory' (Hick, 1997, p. 163) are beyond our scope here (though for a significant critique, see D'Costa, 2009, p. 9–19).

Instead, I wish to highlight something implicit in what has already been stated: the wide range of traditions and sources that AAH proponents appeal to as authentically witnessing (when interpreted correctly) to ancient astronauts interacting with humanity. In this chapter alone, we have noted uses of Christian, Jewish, Muslim, Hindu, Greco-Roman, Egyptian, Sumerian, Native American, Mayan, and Chinese texts and material culture to evidence the AAH. This list is by no means exhaustive; a single episode of *Ancient Aliens*, let alone a single book by von Däniken, could easily add to it. Indeed, a rather large proportion of AAH energy goes into emphasizing similarities across different traditions and cultures: Zeus' thunderbolt and Thor's hammer, for example, or human-and-divine figures such as Greek demigods or the *Nephilim* in Genesis 6 (Ancient Aliens, 2010b). All are taken as, despite their surface differences, really testifying to the same – and seemingly universal – underlying reality.

For AAH theorists, like Hick, the differences can be ascribed to the pre-existing cultural frameworks in which experiences are received, interpreted, and described. (The *locus classicus* here is, again, the terminology of 'chariots' to describe space-age technology: "Wheels and chariots were as close as biblical people came to anything we would call technology. Computers and lasers were not in their lexicon"; Downing [1968] 2019: loc. 351.) Hick's 'transcendent reality', however, is substituted for 'flesh-and-blood extraterrestrials' with advanced technology.

But Is It Atheistic?

The extent to which mainstream AAH is atheistic – meant here in the 'positive' sense of denying the existence of a God or gods (cf. Bullivant, 2013) – is a complicated one and evades any easy answer.

On the face of it, it would appear to be a straightforwardly atheistic, indeed anti-theistic, enterprise. Its primary *modus operandi* consists of taking any and all depictions of, or references to, a God or gods, from seemingly every culture or tradition, and arguing that these God/s are, in fact, 'flesh-and-blood extraterrestrials'. That is, *not* gods at all, except in a metaphorical sense that makes for memorable, and clearly sellable, book titles (e.g., von Däniken's *Gods from Outer Space* [1972a], *The Gold of the Gods* [1972b], *The Arrival of the Gods* [1998], *The Gods Never Left Us* [2017], among others; David Hatcher Childress' *Technology of the Gods* [2000]). Furthermore, as noted earlier, the *assumption* of a naturalist, and thus atheist, worldview constitutes a large part of the argument in favour of the AAH.

Against this, however, one must note that key proponents, including von Däniken, Tsoukalos, and Philip Coppens (2012), have all denied being atheists. Most explicitly, von Däniken has argued, on occasion, for the necessity of a Creator: 'Atheists may deny with ever increasing vehemence the existence of a power whom for want of a better word we call "God." [But:] In the beginning there was a creation'. This ultimate Cause he prefers to refer to as 'IT':

> IT existed before the big bang. IT unleashed the great destruction. IT caused all the worlds in the universe to originate from the explosion. IT, incorporeal primordial force, the decisive primordial command, became matter and IT knew the result of the great explosion. IT wanted to reach the stage of lived experience.
>
> *(1972, chap. 7)*

Strictly speaking, therefore, it seems that the AAH is compatible with affirming the existence of a (real, supernatural) God, alongside all the flesh-and-blood extraterrestrial (metaphorical, natural) 'gods'. Seen in this light, AAH could be viewed as a kind of demythologizing project, clearing away various false conceptions of gods, contradistinguishing these from a 'purer' conception of a true God. In practice, however, this is scarcely the case. One can read a lot of von Däniken's work and never encounter this idea, present though it occasionally is. One could certainly watch dozens, if not hundreds, of episodes of *Ancient Aliens*, without encountering the suggestion that perhaps not every God is a hi-tech alien after all (see also Bullivant, 2023).

This leaves us in an awkward position. On the one hand, the AAH is evidently compatible with theism. It is not, therefore, necessarily an atheistic position. (Incidentally, it holds this is common with other forms of Euhemerism. Several ancient Jewish and Christian authors, for example, advanced Euhemerist interpretations of the surrounding cultures' gods, while affirming the reality of their own one.) However, in practice, it is arguably a significant cultural carrier of atheistic ideas – albeit one that looks and feels very different from other prominent cultural carriers (e.g., works by philosophers and scientists such as Daniel Dennett, Richard Dawkins, or Neil DeGrasse Tyson). It is hard to think of a popular show on American television, for instance, that is so uniformly sceptical in its approach to religious claims as is *Ancient Aliens*. (Bearing in mind, too, that the show, as a rule, never features opposing opinions.) There is, moreover, some evidence of von Däniken's works leading people to embrace atheism (see Brown, 2017, p. 60).

Conclusion

In this chapter, I have attempted to provide a serious philosophical description of a popular package of ideas, which is sometimes termed the Ancient Astronaut Hypothesis. Although most associated with the works of Erich von Däniken, Zecharia Sitchin, and the History channel docuseries *Ancient Aliens*, it also features heavily in science fiction (films, tv, books), as well as in the belief systems of a wide range of New Religious Movements, not least in some of the more (in)famous ones (e.g., Scientology, Heaven's Gate, Raëlism).

It is not, admittedly, a set of ideas often held by, or encountered in the periodicals of, professional philosophers of religion or science. Its inclusion in a serious treatment of 'exophilosophy' might, perhaps, seem a little eccentric. Perhaps so. However, as I have argued, AAH ideas are remarkably prominent out there in the real world – i.e., on the bookshelves, televisions, computer screens, and, at least to some level, in the minds of millions of people. For that reason alone, then as a sociologist of (non)religion and science, it is certainly of interest to me. Millions of people might be wrong, but if so, then they can only be wrong for interesting reasons that are worth understanding. I hope they are of interest to other readers too.

References

Ancient Aliens: The Game (2022). PC [Game]. Los Angeles, CA: Legacy Games.

Ancient Aliens (2010a). The Evidence (History Channel). Season 1, episode 1, first broadcast: 20 April 2010.

Ancient Aliens (2010b). Gods and Aliens (History Channel). Season 2, episode 2, first broadcast: 4 November 2010.

Ancient Aliens (2013a). Magic of the Gods (History Channel). Season 6, episode 4, first broadcast: 21 October 2013.

Ancient Aliens (2013b). Prophets and Prophecies (History Channel). Season 5, episode 7, first broadcast: 8 February 2013.

Ancient Aliens (2014). Aliens and Superheroes (History Channel). Season 8, episode 9, first broadcast: 22 August 2014.

Ancient Aliens (2019). Food of the Gods (History Channel). Season 14, episode 18, first broadcast: 1 November 2019.

Ancient Aliens (2023). The Power of the Obelisks (History Channel). Season 19, episode 4, first broadcast: 3 February 2023.

Asa, Robert (1999). Classic *Star Trek* and the Death of God: A Case Study of "Who Mourns for Adonais?". In Jennifer E. Porter and Darcee L. McLaren, eds., Star Trek *and Sacred Ground: Explorations of* Star Trek, *Religion, and American Culture*. Albany, NY: State University of New York Press, 33–59.

Bader, Christopher D., Joseph O. Baker, and F. Carson Mencken (2017). *Paranormal America: Ghost Encounters, UFO Sightings, Bigfoot Hunts, and Other Curiosities in Religion and Culture*, 2nd ed. New York: New York University Press.

Boyer, Pascal (2001). *Religion Explained: The Evolutionary Origins of Religious Thought.* New York: Basic Books.

Brown, Callum G. (2017). *Becoming Atheist: Humanism and the Secular West.* London: Bloomsbury.

Brown, Truesdell S. (1946). Euhemerus and the Historians. *Harvard Theological Review* 39(4), 259–274.

Bullivant, Stephen (2013). Defining Atheism. In Stephen Bullivant and Michael Ruse, eds., *The Oxford Handbook of Atheism.* Oxford: Oxford University Press, pp. 11–21.

Bullivant, Stephen (2023). Theology Professor: 'Ancient Aliens' Is fantasy Fiction for Atheists. *The Well*, 23 June. Available online: https://bigthink.com/the-well/ancient-aliens-fantasy-fiction-athei sts/ (accessed 18 August 2023).

Bullivant, Stephen, Miguel Farias, Jonathan Lanman, and Lois Lee (2019). *Understanding Unbelief: Atheists and Agnostics around the World: Interim Findings from 2019 Research in Brazil, China, Denmark, Japan, the United Kingdom, and the United States.* Available online: https://resea rch.kent.ac.uk/understandingunbelief/wp-content/uploads/sites/45/2019/05/UUReportRome.pdf

Childress, David Hatcher (2000). *Technology of the Gods: The Incredible Sciences of the Ancients.* Kempton, IL: Unlimited Adventures Press.

Childress, David Hatcher (2016). *Ark of God: The Incredible Power of the Ark of the Covenant.* Kempton, IL: Unlimited Adventures Press.

Clarke, Arthur C. (1953). *Childhood's End.* New York: Ballantine Books.

Clarke, Arthur C. (1968). *2001: A Space Odyssey.* London: Hutchinson.

Colavito, Jason (2020). *The Mound Builder Myth: Fake History and the Hunt for a "Lost White Race".* Norman, OK: University of Oklahoma Press.

Coppens, Philip (2011). *The Ancient Alien Question: A New Inquiry into the Existence, Evidence, and Influence of Ancient Visitors.* Pompton Plains, NJ: New Page Books.

Coppens, Philip (2012). The Atheist Religion, blogpost, 5 February 2012. Available online: http:// philipcoppens.blogspot.com/2012/02/atheist-religion.html (accessed 18 August 2023).

D'Costa, Gavin (2009). *Theology and World Religions: Disputed Questions in the Theology of Religions.* Chichester: Wiley-Blackwell.

Dericquebourg, Régis (2021). Rael and the Raelians. In Benjamin E. Zeller, ed., *Handbook of UFO Religions.* Leiden: Brill, 472–490.

Downing, Barry ([1968] 2019). *The Bible and Flying Saucers: Did a UFO Part the Red Sea?.* Independently published.

Dunn, Christopher (1998). *The Giza Power Plant: Technologies of Ancient Egypt.* Santa Fe, NM: Bear and Company.

Fagan, Garrett G. (ed.) (2006). *Archaeological Fantasies: How Pseudoarchaeology Misrepresents the Past and Misleads the Public.* London: Routledge.

Festinger, Leon, Henry Riecken, and Stanley Schachter (1956). *When Prophecy Fails: A Social and Psychological Study of a Modern Group that Predicted the Destruction of the World.* New York: Harper-Torchbooks.

Fort, Charles (1919). *The Book of the Damned.* New York: Boni and Liveright.

Gallagher, Eugene V. (2010). Extraterrestrial Exegesis: The Raëlian Movement as a Biblical Religion. *Nova Religio*, 14(2), 14–33.

Gordin, Michael D. (2021). *On the Fringe: Where Science Meets Pseudoscience.* New York: Oxford University Press.

Hancock, Graham (2015). *Magicians of the Gods: The Forgotten Wisdom of Earth's Lost Civilisation.* London: Coronet.

Hick, John (1997). The Possibility of Religious Pluralism: A Reply to Gavin D'Costa. *Religious Studies*, 33(2), 161–166.

Hick, John (2004). *An Interpretation of Religion: Human Responses to the Transcendent*, 2nd ed. New Haven, CT: Yale University Press.

Linford, Peter (1999). Deeds of Power: Respect for Religion in *Star Trek: Deep Space Nine*. In Jennifer E. Porter and Darcee L. McLaren, eds., *Star Trek and Sacred Ground: Explorations of Star Trek, Religion, and American Culture*. Albany, NY: State University of New York Press, 77–100.

Lovecraft, H. P. (1928). The Call of Cthulu. Available online: www.hplovecraft.com/writings/texts/fiction/cc.aspx (accessed 25 April 2023).

Palmer, Susan J. (2004). *Aliens Adored: Raël's UFO Religion*. New Brunswick, NJ: Rutgers University Press.

Producers of Ancient Aliens (ed.) (2016). *Ancient Aliens: The Official Companion Book*. San Francisco, CA: HarperOne.

Robbins, Thomas and Philip Charles Lucas (2007). From 'Cults' to New Religious Movements: Coherence, Definition, and Conceptual Framing in the Study of New Religious Movements. In James A. Beckford and N. J. Demerath III, eds., *The SAGE Handbook of the Sociology of Religion*. London: SAGE, 227–247.

Rothstein, Mikael (2021). The Aetherius Society: A Ritual Perspective. In Benjamin E. Zeller, ed., *Handbook of UFO Religions*. Leiden: Brill, 452–471.

Sitchin, Zechariah (1976). *The Twelfth Planet*. New York: Stein and Day.

Sitchin, Zechariah (1990). *Genesis Revisited: Is Modern Science Catching Up with Ancient Knowledge?*. New York: Avon.

Story, Ronald D. (1976). *The Space-Gods Revealed: A Close Look at the Theories of Erich von Däniken*. New York: Barnes and Noble.

Tumminia, Diana G. (2005). *When Prophecy Never Fails: Myth and Reality in a Flying-Saucer Group*. New York: Oxford University Press.

Urban, Hugh B. (2011). *The Church of Scientology: A History of a New Religion*. Princeton, NJ: Princeton University Press.

Urban, Hugh B. (2021). Scientology. In Benjamin E. Zeller, ed., *Handbook of UFO Religions*. Leiden: Brill, pp. 329–342.

von Däniken, Erich (1972a). *Gods from Outer Space*, trans. Michael Heron. New York: Bantam Press.

von Däniken, Erich (1972b). *The Gold of the Gods*, trans. Michael Heron. New York: Putnam's Sons.

von Däniken, Erich (1998). *Arrival of the Gods: Revealing the Alien Landing Sites of Nazca*. Rockport, MA: Element Books.

von Däniken, Erich (2009). *History Is Wrong*, trans. Nicholas Quaintmere. Franklin Lakes, NJ: New Page Books.

von Däniken, Erich (2011). Foreword. In Philip Coppens, ed., *The Ancient Alien Question: A New Inquiry into the Existence, Evidence, and Influence of Ancient Visitors*. Pompton Plains, NJ: New Page Books.

von Däniken, Erich (2017). *The Gods Never Left Us*. Wayne, NJ: New Page Books.

von Däniken, Erich ([1968] 2019). *Chariots of the Gods? Was God an Astronaut?*. London: Souvenir Press.

von Däniken, Erich ([1999] 2019). Foreword from the 1999 Edition. *Chariots of the Gods? Was God an Astronaut?*. London: Souvenir Press.

Zeller, Benjamin E. (2014). *Heaven's Gate: America's UFO Religion*. New York: New York University Press.

ACKNOWLEDGEMENTS

This edited volume emerged out of the 'Religion and Astrobiology in Culture and Society' (RACS) research project, generously supported by a subgrant from the University of Birmingham's 'International Research Network for the Study of Science & Belief in Society' (INSBS), funded by the Templeton Religion Trust.

RACS is a collaboration between St Mary's University (UK), the Lanier Theological Library (USA), Leeds Trinity University (UK), and the University of Notre Dame (Australia). Find our YouTube channel online at: www.youtube.com/@racsnetwork8963

The editor would also like to particularly thank Leeds Trinity University for co-funding the conference that eventually resulted in this volume, as well as Heather Playford and Janet Siefert for their support and encouragement in putting this together.

INDEX

For Product Safety Concerns and Information please contact our EU
representative GPSR@taylorandfrancis.com
Taylor & Francis Verlag GmbH, Kaufingerstraße 24, 80331 München, Germany

www.ingramcontent.com/pod-product-compliance
Lightning Source LLC
Chambersburg PA
CBHW081106220326
41598CB00038B/7250

9 781032 576091